THE ORIGINS OF LIFE

THE
ORIGINS
OF
LIFE

> Evolution as Creation /

HOIMAR v. ⌊DITFURTH ⸲ *Hoimar v.*

Translated by Peter Heinegg

575

1817

Harper & Row, Publishers, San Francisco

Cambridge, Hagerstown, New York, Philadelphia
London, Mexico City, São Paulo, Sydney

Translated from the original German edition, *Wir sind nicht nur von dieser Welt: Naturwissenschaft, Religion und die Zukunft des Menschen* (Hamburg: Hoffman und Campe, 1981).

FIRST EDITION

Designer: Jim Mennick
Drawings by Andrea Schoormanns

Library of Congress Cataloging in Publication Data

Ditfurth, Hoimar von.
THE ORIGINS OF LIFE.

Translation of: Wir sind nicht nur von dieser Welt.
Includes bibliographical references and index.
1. Evolution. 2. Evolution—Religious aspects. 3. Life—Origin. I. Title.
QH366.2.D5613 1982   575.01   82-47757
ISBN 0-06-250909-8

ISBN 0-06-250909-8 (U.S.A. and Canada)
ISBN 0-06-337030-1 (except U.S.A. and Canada)

82 83 84 85 86   10 9 8 7 6 5 4 3 2 1

For Jutta, Christian, Donata, and York

Contents

Introduction
Truth Is Indivisible

CAN WE believe in the existence or, even more, in the active presence of a God in a universe which, after centuries of scientific research, our minds have begun to see as rationally explainable? This simple but absolutely critical question forms the background of this book.

Nowadays the question is seldom posed so directly,[1] which is rather peculiar since at the same time everybody talks—with good reason—about the need to "find meaning" in life. But how can we define convincingly some sort of meaning for our existence without taking a stand on the problem of God? Whatever answer the individual may come up with, it's clear that we can't sensibly discuss the meaning of human existence unless we decide whether this world, the everyday reality surrounding us, is self-contained and intelligible in and of itself—or not.

Yet whether the only reality is this world or there is another world beyond it, as all the major religions have always claimed, is a point that theologians and scientists have long since given up arguing with any seriousness. But not because the issue has been settled. Theologians take the other world for granted (religion *is* the conviction that "the Beyond" really exists), while scientists view it as irrelevant or, at most, something of interest only to the psychologist or sociologist of religion.

There is no open warfare at present between the two camps, but this should not be taken to mean that after centuries of bitter dispute both sides have found the way to common intellectual ground. There is peace, but only because of a compromise: weary of the long conflict, science and theology have simply agreed to proclaim the divisibility of truth.

In the thirteenth century Siger of Brabant taught that what was true for faith could be false for reason, and vice versa.[2] Perhaps he wanted a pretext so that he could speculate freely on philosophical matters in the teeth of theological censorship. (It was all in vain; he was thrown into prison anyway.) But Tertullian (ca. A.D. 160–220) was deadly serious with his defiantly triumphant cry, *"Credo quia absurdum"* (freely translated, "I believe it precisely *because* it seems so unacceptable to my mind"). Whatever sense ancient theologians may have made of Tertullian's remark, modern critics of religion, more cool and objective, would describe it as a typical case of "immunization strategy," because those characterizing their religious standpoint this way have retreated to a position completely beyond the range of rational arguments. They have immunized themselves, so to speak, against every conceivable objection. They have laid claim to a kind of truth unconnected to our notion of the word.

Just as "poetic truth" lays claim to a unique value, although it explicitly distances itself from our everyday concept of truth, so, in the opinion of many theologians, especially Protestants, "religious truth" is radically different from everything that critical reason can judge to be true or false, provable or refutable.[3] Yet while poetic truth doesn't pretend to be anything more than a figure of speech, religious truth lays claim to the full existential meaning of the word.

And so theologians have taken truth apart and shared the pieces with scientists. Evidently they felt this was the only way to avoid the contradictions that were much more deeply feared in the theological camp than on the other side. From then on, one is tempted to say, they kept a nervous eye on the boundary between the two jurisdictions. Whenever we're concerned about the meaning of our lives or worried about our mortality, whenever we wish to measure our behavior against the standards of good and evil, theologians give us the necessary information. On

the other hand, whenever we're interested in the mysteries of the
fixed stars or the structure of matter, in the history of life on earth
or the enigmatic functioning of our brains, we are referred to
another set of truths, the ones in the custody of the sciences.

These two truths, however, have nothing to do with one another—so the theologians tell us, in an effort to reassure themselves and their audience. Thus the two sides keep off each
other's turf. They've stopped competing for each other's clien-
tele. They've gotten to the stage where, by mutual consent, territorial limits are carefully marked off. If nothing else, this undoubtedly avoids a lot of controversy.[4]

The only question is whether theologians can in conscience
excuse themselves this way from carrying on the old quarrel with
science. How can they actually justify their readiness to abandon
"the world" to scientists? How long will theologians ignore the
problems caused by the fact that both truths, scientific and religious, must, when all is said and done, find a place together in
the heads of concrete individuals? How long will they be able to
repress the critical realization that their rapidly increasing loss of
public authority is the inevitable consequence of their failure to
recognize a single truth encompassing both this world and whatever transcends it?[5] How long will they continue to close their
eyes to the fact that they have ceased to take this world seriously
—because if they did, they would also have to take seriously all
our scientific knowledge about the world?

It's one thing to declare that religious truth is completely different from any worldly truth perceived by rational exertion, but
it's quite another to realize concretely in one's life that fear of
death belongs in the selfsame world as nuclear physics, that
moral unrest in the face of certain social structures runs right up
against knowledge of the historical causes of these structures.
Some people expressly declare that the truth they stand for has
nothing, literally nothing at all, to do with the logical and natural
laws that govern this world. Have they any reason to wonder
when their insistence on joining debate over worldly matters
meets with skeptical reserve?

The doctrine of "two truths" saddles us with nothing less than the burden of living in a mentally divided world. In one half we're supposed to believe what we have to reject, on logical grounds, in the other. And, given the incompleteness of the worldly half, we're supposed to orient ourselves by that entirely different truth which, we're confidently told, has nothing in the least to do with the nature of this world. We're supposed to feel responsible for the state of affairs in the earthly part of the world, which nonetheless (our mentors immediately add) in the final analysis doesn't really matter.

All this is more than we can bear. The only reason we don't notice the full absurdity of such mental contortions is the force of habit.

Any unprejudiced observer of this situation must report that the pseudo-solution of the faith/science dilemma has never actually been accepted by anyone, despite all the officially trumpeted agreement. All attempts to talk oneself and others into believing the contrary have come to nothing. They have failed to stifle the unmistakable feeling shared by all those involved that there can be only *one* truth—an indivisible one.

Holding on to a world picture split down the middle has serious consequences. All of us harbor a mute suspicion that only one of the two truths to which we are asked to swear simultaneous allegiance can be true. For the scientist and everyone else whose world view is basically determined by modern science, this has produced a growing tendency to atheism, that is, the attempt to work things out in strictly empirical and rational terms.

But even the theologians don't seem to think the solution they're promoting is all that conclusive. True, the compromise reached has given them some peace from attacks by scientists, but it obviously hasn't quieted their theological conscience. How else can we explain the deep mistrust, the continual and intense uneasiness that large parts of the church (and especially the Catholic church) still display toward science?

As late as 1950, in his encyclical *Humani generis,* Pius XII condemned the proposition that "the origin of the human body from previously existent living matter . . . has already been proved with complete certainty" as an "audacious transgression" of the limits of Catholic intellectual freedom. This defensive attitude unmistakably betrays the fear of a clash between religious truth and some pertinent biological discoveries. Why else issue the warning?

In early 1977 the president of the Commission on Faith of the German Catholic bishops, Cardinal Volk, stated that "the findings of the sciences can be integrated into theology without impairing our faith." Obviously this reassurance, delivered only five years ago, was not superfluous. The Cardinal went to say that this insight gave rise to the "by no means easy task" of articulating theologically the findings of modern science. Thus, as the Cardinal saw it, prior to 1977 this process of articulation had not even begun.

To this day believing Christians, true to the Church, unquestionably find it hard to feel at ease with science, especially with biology. In their heart of hearts they can't get away from the conviction that scientific discoveries relating to creation and living creatures cannot be viewed as something apart from the Creator they believe in. Are there two separate and unrelated truths? Obviously—and fortunately—not even the theologians themselves believe in such a possibility.

So we are right back where we started hundreds of years ago when the question first arose about the relationship between the religious and the rational interpretation of this one cosmos, this one world, which is at the same time creation *and* the object of human scientific study. The question is still unanswered, which need not cause scientists any concern: they have retreated to a methodologically defined area of specialization where they are self-sufficient. Believers have a harder time of it. And it remains a mystery how theologians can manage with the existential schism that marks our present view of the world.

The anxiety that the churches (and, under their influence, so many believers) feel at the prospect of dealing with science prevents them from realizing that for some time now their fears have been groundless. They were apprehensive that an unbridgeable gap might yawn between the world as creation and the world as the object of scientific scrutiny, but the world picture drawn by scientific research in recent years should do away with such apprehensions. Of course, to share in this liberating vision, one has to be willing to take a good, unprejudiced look at it. In any case not since the early Middle Ages have the prospects for harmonizing the religious and scientific views of the world been as bright as they are today. This assertion may at first strike many people as bold, but it can be substantiated, as the rest of the book will show in some detail.

During the Middle Ages Western culture witnessed a tremendous effort, so strenuous that it threw all other questions into the shade, to prove once and for all the existence of God and the reality of a world beyond this one. This led to the conclusion that the task was fundamentally impossible.

Then science entered the scene. Since humans tend to extremes and look for clear-cut, exclusive solutions, the following centuries saw a no less strenuous attempt to refute the existence of God and the Beyond, an attempt to demonstrate that nature and this world can function without the "hypothesis of God," that they are transparent to the human mind. This led once again to the realization that the task was impossible.

Around the beginning of this century science came up with definitive, irrefutable proof that our minds can never fully understand this world. Albert Einstein's theory of relativity may be cited as the decisive instance here.[6] Thus "reason's final conclusion" enables us to see "that there is an infinity of things which are beyond it," just as Pascal predicted more than two hundred years ago in a passage that sounds positively clairvoyant.[7]

This leaves us free to take up once more the problem of synthesizing the religious and scientific views of the world, unburdened now with the prejudices and recrimination of the past. The unnatural split in our sense of self, which has had such serious consequences, no longer seems so irreparable. The recovery of a unified, coherent world picture is near at hand.

The notion that scientific knowledge necessarily contradicts

what religions tells us has turned out to be nothing but a prejudice.[8] But that's not all. Nowadays there is evidence to show that
scientific discoveries and paradigms confirm ancient religious
pronouncements in a surprising fashion.

The purpose of this book is to substantiate that contention. It
was written in the conviction that it has now become possible to
synthesize scientific and religious statements about the world,
and that such a synthesis is not only possible but urgently
needed, if the dwindling credibility of religion is not to sink still
lower.[9] And so I offer this book to the churches as an attempt to
explain, from a scientific perspective, the ways in which science
and religion might after all function harmoniously together and
to sketch out some possible avenues of rapprochement between
the two. Since the author is a scientist, the following presentation
of science's theological implications may quite possibly contain
a few errors, despite all efforts to prevent them. Regrettable as
that would be, it would not invalidate the basic idea behind the
proposal developed here.

At the core of the book is the concept of evolution, which is
applicable far beyond the special case of biology and which is in
fact the fundamental principle of all modern science. Remarkably
enough, this concept provides the key to a better and in many
ways altogether new understanding of some of theology's oldest
assertions, including its faith in the reality of a world beyond our
own. This thought gave the critical impetus to the writing of *The
Origins of Life*, which therefore begins with a relatively long exposition of modern evolutionary theory.

Given the continuing anxieties on the part of believers, it
should be stressed that readers need not fear that in the following
chapters they will be asked to call into question one iota of their
religious conviction. The incorporation of material that science
has brought to light concerning the world and humankind cannot
threaten the stability of theological structures, only bolster them.
However, even though not one supporting wall need be torn
down, renovation cannot be accomplished without a certain
amount of rebuilding.

PART I

Evolution and Belief in Creation

1. Evolution and Human Identity

THE MOST drastic alterations a modern self-critical theologian might be forced to make in his or her faith are due to the fact of evolution. But at this point we have to clear up two misconceptions about the term evolution if the discussion is not to be skewed from the start.

First, evolution is centrally important for our purpose, but *not* because, as some argue, it contradicts the religious interpretation of the world as God's creation. The fact that a large number, perhaps the majority, of so-called educated people apparently believe (and have believed for some time now) that it does makes it necessary to stress, right from the beginning, that this is pure prejudice. But it's a deep-rooted and widespread prejudice, and we shall have to analyze it in some detail in a special section.

Evolution and faith in creation are in no way at variance with each other. (This book, to repeat, was written in the conviction that the world views of science and religion are not mutually exclusive.) On the other hand, their encounter, once it occurs, cannot but have momentous consequences. Thus theologians, if they take seriously their proposed "articulation" of scientific knowledge, very soon discover that many of their formulations are derived from a static medieval world picture that can no longer be considered valid.

At this juncture a single example may clarify the argument. A static world picture operates under the assumption that currently prevailing conditions constitute unalterable reality. It thereby suggests, among other things, either explicitly or implicitly, that the structures and hierarchies which moderns perceive in their environment are the last word and hence the result, accidental or deliberately intended, of the forces or factors that produced this world in the first place. It's a short step from here to the conviction that man is the summit, the "crown" of creation. Proof for this is furnished by the incontestable fact that at the moment humanity stands supreme on this earth.

But seen against the background of evolution things look different. Once we recognize the historical character of nature, indeed of the entire cosmos, we see that the "static" notion of humankind's role in the world leads to the insane conclusion that thirteen or more billion years of cosmic history have served no other purpose than to bring forth modern humanity, with the Cold War and all the host of problems we can only explain as consequences of our highly imperfect nature.

Baldly stated in this way, the assumption that we are the *ne plus ultra* of the universe seems as absurd as it really is. The opposite is true: the discovery of evolution has thrust upon us the absolutely certain realization that the world as we know it cannot be the end (or even the goal) of the cosmic process, that this history will continue to unfold over periods of time as immense as those it has already passed through. From this angle of vision the world of the present—we ourselves and our environment—turns out to be a passing, ephemeral phenomenon. Soberly considered, it is a tiny section, an instantaneous sliver, of a vast ongoing development.

And by the same token the rank we now occupy in the scheme of things is shown to be altogether provisional. (Put aside for the moment the likely existence of intelligent beings on other planets in the cosmos. We shall have to discuss this later on in a somewhat different context.) The mere fact of evolution proves that the role of *Homo sapiens,* as we have come to know it, can be no more permanent or ultimate than that of the Neanderthals or *Homo habilis.* We shall certainly be long gone before the history of the universe has come to an end. We don't know whether we'll simply die out (perhaps in a suicidal holocaust), whether we'll have descendants as genetically distant from us as we are from the mute *Homo habilis,* or whether we might not even constitute a bridge to nonbiological descendants of an entirely different sort.[10]

The one thing we can be sure of is that we are not, if indeed any humans can ever be, the end or goal of evolution. The universe could have gotten along very well without us (as it will one day have to do). Its history would not, for lack of the human race, forfeit its meaning, assuming it has one.[11]

This realization may hurt our pride, but it takes nothing away from our dignity—two things that we ought to be careful to keep separate. But it compels us to rethink the nature of that dignity. And by the way it also restores to Neanderthals and all our other evolutionary precursors the part of *their* dignity which we deprive them of when we reduce them to being our precursors and nothing more. For, borne along as we are by the stream of evolution that encompasses everything, including the future, we are all in the same situation.

The evolutionary point of view further compels us to submit certain religious formulas, especially Christian ones, to critical scrutiny. This evidently holds—to cite, once again, only a single example—for the central Christian concept of God's "Incarnation." There can be no doubt that a Neanderthal would have thought of Jesus not as his "fellow man," but as a divine creature. And there would be the same disproportion, only the other way around, for our future descendants.

There is a clash between the absolute status that Christianity has up till now conferred on the birth in Bethlehem and the identification of the man who was born there with the human race, that is, in the form of *Homo sapiens.* There is general agreement that humanity, even biologically speaking, is imperfect and "unfinished." We have not yet completed the transition from animal to human being, have not yet fully realized our human potential. Theologians and students of behavior can easily present us with a long list of the contradictory, irrational features of human behavior that result from this state of incompleteness.

6 EVOLUTION AND BELIEF IN CREATION

In identifying Christ with such a being can theology really render itself invulnerable, now and forever, to historical relativity? The problem can't be swept away by ignoring the existence of our evolutionary descendants on the grounds that it lies in the future and so has no reality here and now, because "absolute status" implies independence from whatever development the future might bring. It means unchanging, permanent importance ascribed to a concrete historical person, who at the same time must also be thought of as belonging to the species *Homo sapiens.* What is at stake here, as everyone knows, is nothing less than "eternal" truths.

I see no way of removing this contradiction other than to concede that even the person of Jesus Christ has a fundamentally historical and relative character. Actually, why shouldn't this be possible without affecting the substance of faith? The job of working out the proper formula for this problem must be left to the theologians. I can only point out that the problem does in fact exist.

The second misconception that should be mentioned concerns the scope of the term "evolution." For most people it still connotes only the history of biological development, but in the last few decades it has become increasingly clear that the principle of evolution is not limited to the realm of animate nature. It is much more comprehensive than that. It is, to spell it out, the most comprehensive principle imaginable, because it includes the entire universe.

The cosmos is not, as humans believed for millennia, something like an inert container for all the things in the world. It is itself an evolving process that embraces all other modes of development. All of reality, everything in our environment, is essentially historical and developmental. Biological evolution is only a part of the universal process. Hence, if we wish to account for its real significance and understand its laws, we have to take this broad, cosmic framework as our point of departure, which is just what the following chapters will do.

Clearly the fact of evolution must be substantiated before we can turn to its consequences. The foundation for the argument has to be made solid, precisely because those consequences are of such moment and weight. But since prejudice and doubt are still so preponderant in this area—a hundred years after Darwin and two hundred years after Kant, the founder of modern cosmology—some detailed explanation will be necessary.

First, however, I should like, by way of precaution, to repeat once again, for the very last time: this book was written in the conviction that it is possible to harmonize the scientific and the religious interpretations of the world and man. So readers can expect to follow the line of argument in the following chapters without fear of being asked to doubt even a fraction of their religious faith.

2. Cosmic Fossils

EVOLUTION is not limited to the realm of living things—there is also cosmic evolution. And for this reason there are fossils, relics of earlier, long completed stages of development, not only in paleontology (the science of older forms of life) but also in cosmology (the study of the origin, structure, and development of the cosmos). What dinosaur bones are to the paleontologist, helium is to the cosmologist.

Around 7% of all the atoms in the universe are found in the form of helium. This estimate—which, to be sure, holds true only for the parts of the universe reachable by our instruments—is the result of astronomical and especially radioastronomical observations. Why this percentage exactly? Why not half or three times as much?

Remarkably, an answer has just been found to this question. In the last few years cosmologists have put the most advanced computers to work to determine the state of the world in the first moments of its existence.[12] The calculations are extraordinarily difficult, since in the unusually dense fireball—which is what the universe began from—a great number of the most complicated atomic transformations occurred with incredible speed.

The temperature in the ball of fire dropped off rapidly as a result of the ensuing expansion (which took place at the speed of light): from ten trillion degrees a millionth of a second after the world began down to ten billion degrees only a few seconds later. The conditions necessary for the conversion of electrons, positrons, and other elementary particles into one another or into radiation in the form of photons altered from one moment to the next. Meanwhile the rate at which the temperature dropped was in turn variously affected by the interactions among the particles.

Despite the tremendous suddenness of these changes modern computers have made it possible to glimpse for the first time the structures of that primeval chaos, and in so doing they happened upon a situation that cosmologists found extremely interesting. Their calculations showed that a few seconds after the beginning there must have been a phase in which the interplay between the expanding fireball and the processes involving atomic particles kept the temperature for some minutes at just under a billion degrees. During this phase the universe was sufficiently cooled off to allow protons and neutrons to fuse into helium nuclei. But only a few minutes later the temperature of the cosmos dropped beneath the level required for such nuclear fusions (it was now "only" several million degrees), before the nuclei of heavier elements could be formed.

The calculations as a whole showed that the phase of nuclear fusion must have lasted just long enough for about 7% of the available protons to combine with neutrons and be converted into helium nuclei. (In the meantime the remaining protons eventually turned into hydrogen atoms by capturing electrons.) This scenario further dictates that none of heavier elements that follow helium in the periodic table can have emerged during this initial phase of cosmic history. Instead, they were all baked together in the center of the stars, which did not develop until a very long time afterwards out of the primeval hydrogen, and were later released by the explosions of supernovas.

Thus the computer-generated models brought to light what could be called the infancy of the cosmos, whose duration and physical characteristics must have led to the creation of a quantity of helium equal to just 7% of all the matter present in the universe. This figure matches the amount of helium that astronomers find in the cosmos today, roughly 13 or more billion years later—a coincidence most welcome to cosmologists.

In their eyes this makes modern-day helium a remnant, a trace of that developmental phase in the youth of the cosmos. It proves to them the real existence of that unimaginably remote period just as reliably as a bone found today can prove to paleontologists that mastodons lived in the Rhine valley several million years ago. Helium is a cosmic fossil.[13]

If this were a unique case, doubts might still be possible. But even though cosmologists are not so richly blessed with cosmic fossils as their paleontologist colleagues are with skeletons, helium is by no means a special case.

Another and much more graphic example is provided by the so-called globular star clusters. These are spherically shaped star clusters, each one of which consists of from several hundred thousand to millions of suns. Something like 300 of these formations belong to our Milky Way. Their layout and patterns of movement (which are basically different from those of all other suns and sun groups in our galaxy) likewise serve as contemporary evidence that a very long time ago the cosmos must have had a completely different appearance from its present one. Hence the universe as we know it must be the result of an evolutionary, historical process that will run on into the future.

The stars, of course, are neither evenly distributed nor aimlessly scattered throughout the universe. Each one of the countless suns in the cosmos is a building block, which, together with many other suns (up to several hundred billion), goes to make a galaxy, such as our Milky Way. (Analysis of highly sensitive photographic plates shows that there are at least a few hundred billion galaxies in the depths of space, the remotest of which are a billion or more light-years away from us.) A typical example relatively near to us is the famous Andromeda nebula, familiar to us from innumerable reproductions. More than half a century ago, using the largest telescope then available, special photographic plates, and a very long exposure time, astronomers succeeded for the first time in "resolving" its cloudlike semblance and demonstrating that it actually consists of around 200 billion suns.

These are arranged so as to form a flat circular disk, somewhat thicker in the center than on the perimeter, thus looking a little like a discus. The "arms" extending in spirals from the center or axis of the discus to its rim (whence the term spiral nebula) as well as the concentration of almost all the system's suns on a single plane immediately suggest a rotating body.

Nevertheless it took scientists a relatively long time to come up with direct proof that a spiral nebula does in fact turn on its axis in the direction indicated by its arms. Obtaining such proof was made more difficult by, among other things, the extreme slowness (relative to their size) with which these gigantic formations rotate. True, a sun on the periphery of the Andromeda nebula (or of the Milky Way, to which our sun belongs) traverses no fewer than 500 kilometers per second. But because of the enormous circumference of a galactic system (whose diameter may reach 100,000 light-years) a single complete revolution requires so much time that the fastest moving galaxies have only managed to turn about their axes 20 times since their origin ten billion years or so ago.

The "fossilized" character of the globular star clusters consists in the fact that they are the only sun groups in the entire system that do not join in the general rotation. Needless to say, they do revolve around the center of the galaxy, which is their common center of gravity; otherwise, with no centrifugal force to stop them, they would plunge into this center. But they, and they alone, behave like individualists, avoiding the familiar route around the disk taken by all the other suns and sun groups. Nor is there any coordination linking the movement of one star cluster to another. Every one of the 300 or so star clusters in the Milky Way follows an orbit on a different plane and in a different direction from those of all other members of the galaxy. Together they fill the space that surrounds the disk of the Milky Way in the form of a sphere.

The explanation for this oddly independent behavior is not hard to find. The spherical space in which the globular star clusters revolve (the "halo") obviously corresponds to the space formerly occupied by the matter in our galaxy when, at the beginning of its existence, it was still a turbulent cloud of gas. When

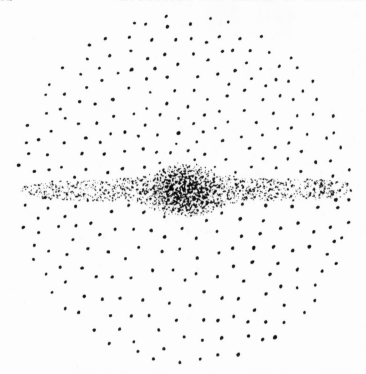

Average distribution of the star clusters belonging to the
Milky Way. The disk-shaped galaxy is made up of about 200
billion stars: Here it is shown from a side angle with the star
clusters surrounding it in the form of a spherical "halo."

this cloud, under the influence of its own gravitational attraction,
began to collapse into itself, it took on a spinning motion, at first
extremely slow, then increasingly faster. Like an ice skater, tightly
pressing his arms against his body when making a spin, the galactic cloud revolved with ever greater speed as its diameter contracted. Its originally spherical shape steadily flattened out into
a disk as a result of the increasing centrifugal force. Finally a state
of equilibrium was reached in which gravitational attraction bal-
anced centrifugal force. Since then the disk has gone on rotating
at a steady velocity while maintaining the same shape.

But the effects of the giant cloud's inner attractive forces were not restricted to the cloud's general structure. At numerous points local centers of gravity also came into being, whose gravitational forces led to the concentration of matter. These local condensations were the source of the first stars. And once again this process began slowly but then continually picked up speed. It's easy to see how more and more local concentrations of matter would form "seed crystals" for the creation of stars as the cloud went on thickening.

The whole process took place with notable rapidity. Specialists estimate that from the initial contraction of the original cloud until the emergence of the complete Milky Way a period of only a few hundred million years elapsed. For astronomers that's a relatively short time.

Yet how do we know that events must have followed the pattern sketched out here? For one thing, on the basis of computerized models. And how can we be sure that this scenario, certified by the computer as physically and mathematically possible but dealing with things that happened some ten billion years ago, corresponds to reality? The answer is (among other things) from the behavior of the globular clusters.

There's no other way to understand it. Quite obviously, the stars in these clusters were formed together in a very early phase of the process described above, at a time when the galactic cloud had not even begun to rotate. How else can we explain the fact that the clusters containing these stars do not lie in the plane of the present day Milky Way and are not subject to its direction of rotation? Or the fact that each of these clusters pursues its own individual direction of rotation independently of the others? Their paths were determined by the influence of accidental local centers of gravity and proper motion long before the system as a whole had started to move and had imposed a common direction on all its elements.

That this interpretation is correct is confirmed by a further discovery. The members of the globular star clusters are by a wide margin the oldest stars we know of. Their age has been estimated on the basis of various kinds of evidence to be at least 13 billion years; so they obviously date from the birth of the Milky

Way. In the case of all other galaxies close enough to us to permit investigation of *their* surrounding globular star clusters, exactly the same results have been obtained.

Hence there can be no doubt that the cosmos did not always present the same appearance that it does to us today. It has "developed," is developing at this very moment, and will go on developing into the future. The only reason it took the most strenuous efforts of human intelligence to discover this fact is that our life span, measured against cosmic processes, is so short. Needless to say, there have been no conspicuous changes during the few millennia that humans have given any thought to the real nature of the starry sky they watched night after night. Nevertheless not the slightest doubt remains that the universe is constantly changing and evolving. There is no need to adduce further proofs of this. So much has been written about the modern picture of the cosmos that most people are familiar with it, at least in broad outline, and have readily accepted it without hesitation.

The same is true for the idea of planetary evolution. Nobody has any problems with the fact that the solar system is not eternal, that the earth and the other planets only came into existence about five billion years ago. People easily integrate this into their notion of the universe; they raise no objections to the thought that the world and nature develop. I'm not aware that anyone has been scandalized by the history of the earth's crust as reconstructed by physicists and geologists.

As far as I can see, nobody is disturbed by this chapter of evolutionary history. Science claims that on the surface of our planet all the physical and chemical factors that were the absolutely indispensable prerequisite for the transition to the biological stage of evolution finally came together. No difficulties here.

But the situation undergoes an abrupt and radical change when the next step in the evolutionary process comes up. Many people, as we know, find it very hard to think of the first life forms as a continuation of the previous stages. Religious people especially are provoked by the assertion that biogenesis and the biological phase of universal evolution which begins with it are to be explained by the same natural laws that govern all the earlier phases. Religious conviction runs headlong into the scientific pronouncement that the emergence of life and the course of *biological* evolution was just as "natural" as the first chapters of cosmic evolution.

At this point—but not before—they think they're obliged to see a fundamental break with the past. The world and nature suddenly divide in half: one side is controlled by the laws of nature, while the other cannot be explained and understood by these laws alone, but only by an additional feature, the immediate, transcendent intervention of a divine creator.

This sort of attitude is the very simple reason why in this book cosmic evolution can be briefly dealt with, while biological evolution will have to be discussed at some length. Our subject is not the natural history of the world.[14] but the problem of harmonizing religious and scientific truth. Under the circumstances we have to pay special attention to the stumbling blocks in this area.

The greatest of these is undoubtedly biological evolution, and hence it will be treated first. We shall have to follow a definite order here: to begin with, we need to spell out the evidence and arguments that establish the fact of phylogeny (the evolution of species) and the regularity of its historical course. Then the most serious objections to evolution must be considered and critically tested. Only after that can we turn to the decisive question of whether and in what sense the theory of evolution and belief in creation are compatible.

But here I want to anticipate and present, in outline form at least, just one of the arguments that have to be thoroughly discussed in the final chapter of this part, which is the linchpin of the book. I do this in the hope of giving skeptics something to think about right from the start. It may dispose a few such people to take notice of the arguments put forth in the following sections, instead of immediately brushing them aside.

Let me clothe this argument in a question: do the people who believe in God (I'm one of them) and who think their faith obliges them to reject on principle the possibility that life can be explained by the laws of nature, do these people see that in so doing they truncate the creation of the God they believe in by removing from it a part of the world and of nature?

If I admit that the evolution of the cosmos, of all inanimate things, is a natural process, only to deny in the next breath that the same holds for life on the grounds that its "creaturely character" demands a supernatural cause, then I am also tacitly saying that sun, moon, and stars manifestly do *not* belong to God's creation, at any rate not in same sense that living "creatures" do. Such a conclusion ought to make one suspicious.

But I'll leave this suggestion at that. We have to take things in order, and so finally let us proceed to the facts of biological evolution.

Three issues have to be distinguished here. First, we have to support the assertion that there is such a thing as phylogenetic history; that the species of plants and animals now living on the earth have not been there from the beginning—nor did they come into existence simultaneously through a single, unique act of creation; that instead they were the result of an extremely long process which has led from the simplest primitive forms of life through countless intermediate steps to the higher animals we know today and to the human race.

After dealing with the reality of biological evolution we then have to look into the problem of its beginning. Thus our second question addresses the classic enigma of "spontaneous generation," that is, how to explain scientifically the emergence of the first life forms from inanimate matter.

Finally, the third problem concerns the *laws* of evolution. If phylogenesis is a biological fact set in motion by the appearance of life (which took place under perfectly natural conditions), the question still remains whether the course of this development, which led in an apparently *purposeful* manner to ever more complex and "higher" forms of life, can still be understood as the product of nature's laws.

3. The Reality of Phylogenesis

TRUE ENOUGH, nobody has ever seen biological evolution going on right before his eyes. But that's no reason for irritation. By the same token no human eye could ever have witnessed the origin of the Milky Way. Nevertheless, in view of the evidence left behind by various "fossils," we have no rational grounds for doubting either the fact that presently observable galaxies are the result of cosmic evolutionary processes or the manner in which this occurred. The same is true of biological evolution.

The parallel here, to be sure, consists simply in the fact that events which can't be directly perceived (because, say, of immense differences in "timetables") can still be proved empirically, if indirectly. The fossils, however, in the concrete, original sense of the word, are less useful as proof, with reference to our first question, than some people might think because those who to this day deny the *fact* of evolution (henceforth I shall use the word without any adjective to mean *biological* evolution) do not look on the many ancient skeletons that have been unearthed as incontestable evidence.

To begin with, a word on the importance of these dissenters, whose opinions are grossly overvalued by many laypeople. Nowadays the overwhelming majority even of the opponents of Darwin's theory (i.e., that the concrete paths taken by evolutionary history are explained by natural laws) admit that evolution itself is a fact, that in the last four billion years biological development *has* taken place on the surface of the earth. But at this late date there are still eccentric loners who deny that evolution ever occurred. Among these every now and again one even finds writers sporting academic titles.

As a science editor I regularly get copies of "anti-Darwinian" articles in the mail. The covering letters declare, usually in a reproachful tone, that an author from the opposition has obviously been overlooked and that as a result the existence of contrary opinions has been hushed up. As laity, the people who send such letters have no way of knowing that every single one of the anti-Darwinists (with scientific pretensions), even the professors and Ph.D.s, are outsiders to the scholarly community. This says nothing, in principle, against the possibility that a dwindling minority could be right on occasion. Hence this book, which is aimed at the general public and not at specialists, will look into the arguments and objections raised by the anti-Darwinian prophets.

The mere fact that such people exist proves nothing at all. There is no idea, however insane, for which a few individuals have not been willing to take the stand. In recent years ordained ministers have proclaimed their atheism from the pulpit (and then naively expressed outrage upon being removed). Would this justify my claiming that the question of God's existence is evidently "still being disputed in the Church itself"?

In England, the home of cranks and nonconformists, there is a club still in operation whose members deem it their responsibility to promote the view that the world is not round but flat. And there are supposedly university men on its rolls. Does this therefore oblige all geographers to regard the roundness of the earth as "controversial"? Objectively considered, these examples give a fair sense of the number and rank (as judged by their colleagues) of the sort of scientists who dispute the fact of evolution or argue that the theory of evolution is inconclusive.

In such situations laity often point out that this might just be one of those cases where the "scientific Establishment"—as it has done in the past—either can't or won't recognize a new idea precisely because of its newness. But this argument doesn't work, for various reasons.

First of all, the dismissal of evolution is no "new idea," but just the opposite: the attempt to reject a new idea and to turn back the clock of knowledge. This may be appropriate in some cases, but needless to say it doesn't in itself justify the appeal to the rejection of brilliant ideas in the past.

Second, such episodes may have actually happened, but much less often than the biased observer would admit. There is no convincing example of a truly revolutionary, brilliant idea ever falling victim to a lack of understanding by scientific "doctrinaires." There has always been resistance to new theories, but it has never won the day. Darwin, we know, did not meet with hostility from his fellow scientists, and neither Copernicus nor Galileo was opposed by the astronomers of their epoch. The opposition, it may be remembered, came from a different quarter.

Furthermore, even if there were cases in which the "scientific Establishment" did succeed in temporarily stifling a new idea, this would not validate the reverse proposition that every idea rejected by the Establishment must therefore be revolutionary— or even of any value at all. Rare exceptions aside, rejection takes place because the idea in question is purely and simply wrong.

But back to the issue of the proofs for biological evolution. As we said, fossils (petrified remnants of extinct life forms) do not qualify as arguments in its favor. At least not for persons who insist that mere probability, however overwhelming, is not enough, who are bent on defending their position come hell or high water, no matter how farfetched a case they have to make. In fact, a strict "creationist" will never, regardless of fossil discoveries, deviate from the conviction that God simultaneously created all existing species of plants and animals in the period of time described by Genesis.[15]

What do the many finds of dinosaur bones indicate? The skeptic could cheerfully concede that these reptiles once ruled the earth and then became extinct a long, long time ago. He would merely challenge the claim that they were not present from the onset of creation and that they could be the biological ancestors of mammals or birds.

These skeptics would likewise be forced to acknowledge the existence of extinct anthropoids such as Neanderthals or *Homo habilis,* but they would not have to deny them to maintain a consistent creationism. Once again they would only need to contest the evolutionary change leading from one of these prehuman forms to another, that is, the relationship between anthropoidal ancestors and their human descendants.

This is, as I see it, the true underlying reason for the continuing resistance to evolution, which can only be explained psychologically since it has long ceased to have any basis in logic: creationists realize that once they accept the idea of evolution, they inevitably have to bring humans, that is, themselves, into the process. Even today this very notion triggers in the minds of many people prejudices and misconceptions so powerful that we shall devote an entire chapter to them.

But what argument could be strong enough to get a dyed-in-the-wool creationist thinking seriously on these matters? The only one that should by rights be effective is based on the unquestionable connection between resemblance and relationship. We take this connection for granted and consider it self-explanatory when it comes to the members of our own family. We expect children to look like their parents. We feel gratified to note the resemblance between a child and its father, because this provides undeniable evidence for all eyes to see that the two are related by blood.

We also take it for granted that more distant relationships should be marked by fainter degrees of resemblance. When we leaf through old family photo albums, none of us ever doubts that the similarities between members of different generations, as seen in the pictures, derive from the fact that these people belong to the same family not merely in the legal but in the biological sense, and that hence these similarities are proof of a genetic relationship. We assume that such individuals have a common family tree, that is, are related by genealogical succession.

Even a creationist, one imagines, would have a hard time forgetting all this when looking at the tremendous similarities between human and nonhuman (whether animal or plant) life forms

on the earth. The fact that an ape has two eyes and two ears, five
fingers on each hand and the same number of vertebrae as we do
cannot—unless we refuse to admit arguments based on probability—be explained as mere coincidence any more than the facial
resemblance of two cousins can.

"Just think, my dear, they tell us we're related to the monkeys.
Let's hope it isn't true. But if it is, then let us pray that the word
doesn't get around." These remarks were supposedly made by
the wife of a prominent British clergyman to a friend around the
turn of the century, after she had wandered into a lecture and
heard for the first time about Darwinism.

The shock and alarm that the lady felt links her to modern-day
creationists, but she was far superior to them in critical intelligence. For while the creationist retreats to the position that what
(in his opinion) ought not to be cannot be, the English lady never
doubted, even in her first pangs of dismay, that the question
whether the shocking discovery was true or not could in no way
depend on her own hopes or fears.

All mammals have seven cervical vertebrae, from the stubby-
necked mole to the giraffe, which owes its proverbially long neck
not to an increased number of vertebrae but to their elongation.
How else can we explain this except by descent from common
ancestors? How else explain the five projections on the forelimb
found not only in the human hand but also in the wings of a bat,
the shovel-like paw of a mole, and the fin of a whale?

"Chance, pure chance" is the cry from the creationist camp,
where they seem to look down upon statistical probability. As
outrageously improbable as the explanation through sheer coincidence may be, people do employ it, and for this reason we shall
have to go into some details of molecular biology. There we shall
encounter patterns of similarity in which the numbers that come
into play are such that they banish the possibility of a simply
accidental congruence.

There wouldn't be a single creationist left if they didn't believe
they could dodge even mathematical proof with one final subter-
fuge. But this, as we shall see, is so preposterous that it needs no
refutation.

4. In Search of a Molecular Fossil

A PETRIFIED bone or its imprint is the most familiar kind of fossil but by no means the only kind. We saw this in the discussion of the peculiar behavior of globular star clusters. A fossil is simply the trace of an earlier stage of development. Thus there are also biochemical and molecular fossils.

Animate nature is no less conservative than the earth's crust. It too preserves the vestiges of its past for incredibly long periods of time. The reason for this is closely bound up with its basic mode of operation: every one of the crucial "inventions" that life has made on earth constitutes an achievement in the face of such high odds, a stroke of such extraordinary luck that nature clings to it with all the obstinacy that the mechanism called heredity puts at her disposal.

Heredity can be thought of as a kind of memory, a storeroom for all of life's successful innovations, the victories it has won. The defeats are forgotten (no mistakes are kept on record in the genetic code). This is why nature learns nothing from her mistakes. Like an unteachable child she continually repeats them, paying no heed to her many past failures.

This comparison is not as lame as comparisons usually are: nature, or evolution, is incapable of learning by experience how to vary her strategies in accordance with her rate of success. So, for example, she keeps on producing albinos in countless animal species without the slightest chance of ever learning that an albino blackbird, a white deer, and a white mouse, because of their conspicuousness in the wild, are "underprivileged" with respect to the most important natural quality of all: the ability to survive.[16]

While evolution doesn't learn from its mistakes, it holds on to its "lucky hits" with a doggedness that persists over billions of years. The mechanism that makes this possible is heredity, the molecular apparatus of the genetic code, whose submicroscopic structure we have already begun to understand in part. "Heredity" means simply that life is spared the trouble of having to reinvent in each generation all the equipment needed for the survival of an organism. That, as a matter of fact, would be an impossible task, a truly lethal precondition. Life has managed to exist on earth for so long only because it has been dispensed from this requirement, which is taken care of by the genetic transmission device known as heredity. This stores up all the functions and structures necessary for life in the form of a code and passes them on from generation to generation. Heredity is nothing but evolution's memory or, to put it another way, genetic tradition.

Now—and with this point we come closer to the heart of the matter—the sequence in which the structures and functions that make up the "blueprint" of a living creature were inserted into this genetic code had to be governed by a definite priority. If a given species has passed through an evolutionary history spanning millions of years and generations, then we have to expect that it stored up the basic functions guaranteeing its members' survival *before* acquiring any special features which may be characteristic of the species in its present-day form. It's really a truism: before procuring the luxury of wings, antennae, or other special accessories, first you have to make sure of the elementary vital functions, such as metabolic processes. The "fundamentals" of life had to be worked out before any features that were not strictly necessary.

Again, this is perfectly obvious, but it leads to an inference crucial to our argument. There are, without a doubt, life functions so elementary that their presence can be demonstrated in *all* living creatures. These include, for instance, the metabolic functions required for the assimilation and conversion of energy. A living being that does not engage in a continual exchange of energy with its environment is simply unthinkable.

Hence if it were possible to find specific molecular vehicles for certain metabolic processes, this would enable us to make a prediction that, once tested, ought to decide once and for all the issue of whether biological evolution occurred on this earth or not. For if such specific vehicles do exist for elementary biological processes, then if evolution *had* occurred they would have to show up identically in all presently existing life forms, from amoebas to elephants. This is simply because they are responsible for vital functions of such an elementary sort that they would have had to come into being at a time before the descendants of the "primal cell" underwent the evolutionary split into the multiplicity of modern-day species.

As a matter of fact, such specific vehicles do exist. They are called enzymes—complex molecules that, depending upon their individual structure, trigger various steps in the metabolic process. And they can be found in all life forms without exception that have been studied so far—from the elephant to the amoeba and from the human being to the grain of wheat in the field. Their complex structure makes each one of these enzymes such an unmistakable "individual molecule" that they can be identified with absolute certainty in the most varied organisms. In other words, we have here yet another kind of "resemblance," this time on the molecular level. The fact that this molecular similarity evidently links all forms of life together proves that they are all interrelated. They all belong to a single family tree that embraces all earthly life. Evolution is a reality.

We can't, of course, simply make this claim and leave it at that. Its importance for this book is so great that we need to carry the argument through by examining a suitable enzyme molecule. I shall take as my example an enzyme that scientists call cytochrome C.[17] We shall see that the fossil-like nature of this molecule arises precisely from the minimal differences which can be ascertained among its copies in the different species of plants of animals—however paradoxical that may sound after all that has been said here.

We know that cytochrome C has to be extremely old from its elementary function (intracellular oxidation). It carries out this very function in all forms of life in accordance with the same principle, which proves that all these forms are bound together in a universal relationship, derived as they are from a single common ancestor, a "primal cell." A long time ago that cell hit upon intracellular oxidation, and then the hereditary mechanism passed it on to all its descendants. But the minimal variations that have been detected in the cytochrome C molecule (depending upon which species it comes from) make it possible to trace the course of biological history. They allow us to reconstruct the temporal rhythm of evolution together with the points on the family tree of life where the various branches forked off and the times when this took place.

5. The History of Cytochrome C

IN EVERY one of the roughly 50 billion cells that make up the human body there are hundreds of complicated chemical reactions that must occur each second if we are to stay alive. They must all run their course within the most narrow confines, at a temperature of only slightly more than 37° C, at speeds that are sometimes on the order of a fraction of 1/10,000th of a second, and in such a way that none of the reactions, which follow in such blindingly fast succession, interferes with any of the others. Life as we know it presupposes that these difficulties have been resolved within the microcosm that is each cell.

These problems seem almost insoluble, but nature has managed to overcome them in a way that, relative to their complexity, strikes us as both simple and ingenious. She has developed "molecular keys," which scientists call enzymes. These are molecules with such an "unlikely" shape that they can be compared in both function and appearance to safety keys.

The reliability of a key, the security it gives to its owner and legitimate user, depends on the degree of its specificity: that is, to what extent the lock for which it has been made can be opened with other ("alien") keys. The probability of this happening with primitive locks is great. It is highly probable that the bit of a key, consisting only of a simple metal square, to a medieval household chest would also fit the locks of many cabinets built in the same period. This key then is largely "nonspecific" and so provides only minimal security.

The probability that a lock can be opened with any given key decreases, for obvious reasons, in proportion as the bit required to open it becomes more complicated. To sum up, the specificity of a key is "negatively correlated" with the chance that there might accidentally be a second key with the same bit. The smaller this chance, the greater the key's specificity (and security). The *unusualness* of the key bit is the index of its specificity, of the security it furnishes the lock for which it was made.

The most secure (or most specific) key imaginable would undoubtedly be a key with a bit that was absolutely singular and unique in the world. This is why in the future security locks will most likely be developed that will respond by means of an electronic recognition device when the authorized person—and no one else—presses it with his or her thumb. When our own fingerprint replaces the bit, then, as we have learned from dactylography and criminology, we shall have the perfect key.

Whether there are any mechanical keys nowadays that meet this ideal requirement is questionable—but, practically speaking, it's also immaterial, because there is no danger to a locked safe merely because somewhere in the world there may be a few dozen people who unknowingly have keys that would open it. But in the molecular sphere such "absolutely specific" keys do exist and have for hundreds of millions of years. These are the enzymes. Let's begin by taking a closer look at the structure of these molecules, which are responsible for certain chemical reactions within the cell, in order to understand why and in what sense they are rightly regarded as keys.

Enzymes are proteins and, like all proteins, are made up of amino acids. We need not be concerned here with the composition of an amino acid (a nitrogenous organic acid). But it's important to know that of the hundreds (or more) of amino acids that a chemist could think up or manufacture in the laboratory, only 20 of these are to be found in living creatures. All the thousands and thousands of different proteins that may be present in any cell existing on this earth are put together with these same recurrent 20 amino acids as elements or building blocks.

But they would be better compared to a string of pearls than to building blocks, since all proteins are strings of molecules—chainlike sequences of the same 20 amino acids in ever changing order. The length of the chain varies from protein to protein. Insulin, for example, is a hormone consisting of 51 amino acids (the same 20 again, but each one can appear more than once at various points on the molecular chain.)

The enzyme cytochrome C is a molecular string composed of 104 amino acids. Its structure is schematically depicted on page 32. Each of its constituent amino acids is characterized by a different printed symbol. This sketch is a concrete representation of the precise order in which these amino acids are arranged in the molecule that the biochemists have dubbed cytochrome C. As we have already noted, cytochrome C is an enzyme, a "metabolic key."

In contrast to an ordinary key, the specificity or certainty with which cytochrome C triggers (or unlocks) one particular reaction can be quantified. Specificity, it may be recalled, equals the degree of probability that a form identical to that of the key might by pure chance exist in another place. But in the case of cytochrome C this probability is nil.

This can be easily proved. The likelihood of a purely accidental "repetition" here is equivalent to the concrete chance that cytochrome C's characteristic sequential pattern of amino acids, which make up its "string," could come about by accident. Putting it more simply: how often would one have to throw 104 pearls of the proper 20 colors (corresponding to the 104 amino acid building blocks of the enzyme) into a groove before they matched, by pure chance, the exact original order?

The question can be answered precisely. The number of possible arrangements of 20 different elements in a chain made up of 104 links is exactly 20^{104}. The order followed in the cytochrome C is just one of these. The probability of reproducing this order by pure chance amounts therefore to 1 in 20^{104} or, converting to the more usual base 10, 1 in 10^{130}. Thus it is clear that the enzyme-key cytochrome C cannot have come into existence a second time through pure chance either on earth or anywhere else in the cosmos.

Since the beginning of the world, since the "Big Bang," only 10^{17} seconds have passed (that's how large exponential numbers are). So if once a second since then the 104 "pearls" had been thrown out at random, this would have produced at the most 10^{17} variations of the string. And the exact sequence of cytochrome C would certainly not have been among them. Even if every single one of the molecules in the entire cosmos represented a different variant of the 104-member string, even then we can be practically certain that the whole cosmos would not contain a single molecule of cytochrome C, because there are only about 10^{80} atoms in the universe.

Thus we have to admit that the possibility of a purely accidental repetition of this special molecular sequence of amino acids on the relatively confined area of the earth's surface can, under these circumstances, be excluded with reasonable certainty. Yet we keep running into this sequence again and again on the earth, namely in every living creature that has been studied so far: not only in humans, but in monkeys, dogs, and ants, in fish, frogs, and butterflies, as well as in mildew, wheat, and ordinary baker's yeast.

How is this possible, if we can confidently rule out chance as the cause of such a remarkable coincidence? Given the circumstances, how else can we explain the fact that they share the same highly specific enzyme pattern except by a *relationship* that binds all living creatures together? And how can this relationship be understood except as a community of organisms that has arisen thanks to heredity, or "genetic tradition"? Despite all the differ-

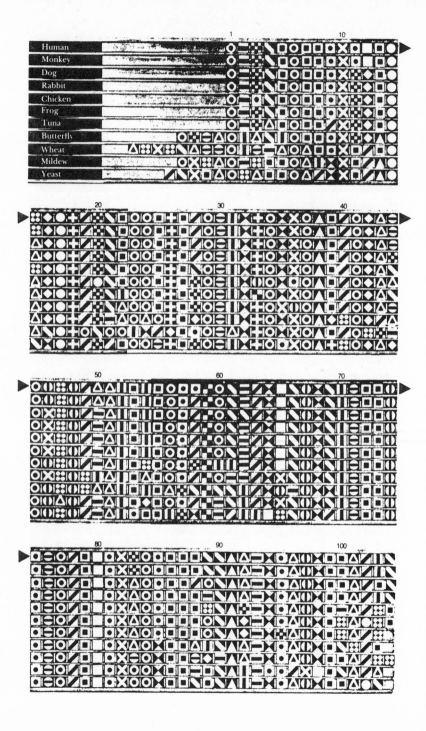

ences separating them, aren't we forced to regard these organisms as the descendants of one and the same primal cell?[18]

In their determination to challenge the reality of evolution (because they erroneously presume that if they don't their religious faith will suffer) creationists have a ready answer. It goes like this: obviously chance is not the explanation, but there's no proof of a genetic connection, either. God simply arranged things during creation so that all the living beings he fashioned were provided with these and countless other identical enzymes.

But this objection can be dealt with too. Here is where the minimal "deviations" previously mentioned come into play—the amino acid sequences of cytochrome C detected in the various species are actually not absolutely identical. In the illustration on page 49a the sequences for humans and for ten other species have been listed together for purposes of comparison. The order followed corresponds to diminishing levels of relationship.

This comparison reveals a further and extraordinarily interesting fact: the number of differences rises as we go from top to bottom. Between the amino acid sequence of cytochrome C in humans and in the rhesus monkey (line 2) there is only a single difference (at position 58). But when we compare the human pattern with that of a dog, we discover 11 variations, and so on down. The more distant the connection, the greater the number of differences.

Strictly speaking, then, in our earlier calculation of probabilities, instead of the number 20^{104}, which holds for the different human races, lower exponents would have to be substituted, depending upon the species being compared with humans. In the case of the rhesus monkey the figure would be 20^{103}, for the dog 20^{93}, and so forth. But these corrections have no substantial effect on our argument, especially if we don't limit it to humans but compare the various species with one another. Even then the resulting probabilities remain so extremely small that the possibility of explaining similarities by pure chance can henceforth be ruled out.

But what is the cause of these individual variations that an originally identical molecule has gone through? We can get closer to the answer by keeping in mind how many times this molecule must have been reproduced ("copied") ever since it emerged from the evolutionary process in the unimaginably distant past. Every time a cell divided the molecular blueprint had to be duplicated in all its particulars. And then in each new cell the work of synthesis had to begin all over again with the help of this copied blueprint.

Now the molecular-biological apparatus in charge of this hereditary process functions with an almost incredible degree of precision. But it's not absolutely perfect. It can't perform impeccably, given the number of atoms that it has to arrange in a certain pattern each second and given the mind-boggling periods of time over which it has to do this, each and every instant. There's no such thing as absolute perfection in the cosmos. Not on the molecular level certainly—the natural radioactivity of the environment and the disturbances from inevitable thermal movement make that impossible.

So mistakes do creep in. They're astonishingly infrequent, if we consider the delicacy of the processes of molecular reproduction, but over the course of time they add up. Then we suddenly find, after the molecule has been reshaped in a newly formed cell, a link of the "wrong color" at a specific point of the chain. One amino acid has been exchanged for another; an alteration of hereditary material has occurred which biologists call a mutation.

Whether this exchange will have any consequences depends upon the point on the molecular chain where the change in sequence has occurred. If the mutation affects the so-called active center of the molecule, then the results with a few rare exceptions are fatal. The active center is something like the bit of the enzyme key, and thus the part of the molecule crucial to its metabolic functioning. In the case of cytochrome C this consists, as we have seen, in oxidation within the cell. Hence a mutation in the area of cytochrome C's active center generally leads to the immediate death of the newly developed cell owing to internal asphyxiation: the mutation has proved lethal.

The rare exception to this rule would be the situation where the new amino acid suddenly appearing in the active center led to an *improvement* in the molecule's efficiency; that is, if its emergence should, quite by accident, alter the molecule's "bit" so as to better the process of oxidation, by making it faster, say, or more effective. We can readily understand that this occurs with extraordinary infrequency. On the other hand when it does, every such case constitutes a "success" that evolution will stubbornly cling to thereafter.

This takes place thanks to the simple fact that an organism which has undergone an improvement by winning the mutation lottery has a better chance of survival. Hence it will very likely leave behind a greater number of descendants which, since mutations are hereditary, will be provided with the same advantage. The new type will therefore prevail, even numerically, over the less well equipped competitors from its own species before many generations have passed.

Biologists are now agreed that this is how the specificity of an enzyme comes into being over the course of long periods of time: every "lucky hit" was retained, while all cells with a counterproductive mutation within the active center (that would undoubtedly be the vast majority) died off. They were eliminated and all traces of them have been expunged.

But in addition to these there were also mutations that had no effect on the active center of the molecule, but only on the "grip" of the key, the purely static part of the enzymatic molecular structure. These are obviously the cause of the variations we find today among cytochrome C molecules from different sources, whether from humans, insects, or plants. The evidence for lethal mutations no longer exists. The bearers of lethal mutations could not survive (and thus could not introduce their negative qualities into "genetic tradition.") The extremely rare positive mutations led to the specificity we can now observe and to the effectiveness of "modern" cytochrome C, which could hardly be surpassed. But all mutations taking place elsewhere in the molecule gradually managed to accumulate over the course of time, since they were immaterial to the enzyme's function.

"Gradually," "over the course of time," are words that have to be stressed. The more time passed, the more the number grew of amino acids replaced by mutation at functionally neutral points in the molecular structure. This had no effect on the enzyme's functioning, which is the only reason why these increasing (by degrees) differences could be transmitted in the first place. But in a surprising turn their interrelations prove to be an evolutionary calendar of remarkable accuracy.

The greater the number of differences, the longer the interval of time. If evolution does exist, this casually drawn conclusion allows us to take two different types of currently available cytochrome C and to estimate how far back in the past we should place the common progenitor of both organisms which contain the different molecules.

Within a single species an exchange of amino acids at intervals averaging, say, hundreds of thousands of years will (sooner or later) infallibly become a distinctive feature shared by all the members. A species is a procreative community. The thorough sexual mixing of the species' gene pool, the totality of its typical genetic traits, assures that these traits will be apportioned throughout the entire population.

But when a new species begins to evolve, that is, when a genetically novel type splits off from what till then had been the common line of descent, this means that the new form has left the hereditary community.[19] From now on the new species will have its own individual genetic fate. In other words the accidental mutations that will continue to occur as time goes on at the most varied points of the cytochrome C structure *will no longer be exchanged* across the new boundary line now separating the two species.

Once the evolutionary break has taken place, therefore, the accidental alterations at functionally insignificant points of the enzyme in both species will proceed to develop independently of one another. And precisely because the accidental copying errors that cause them gradually grow more numerous "over the course of time," *the present number of mutations serves to measure the time elapsed since the two species split apart.*

Unfortunately, in the cold light of the laboratory reading the evolutionary calendar on the basis of these figures proves to be somewhat harder and less reliable than the relatively simple principle just enounced might suggest. Merely counting the differences in the amino acid sequences is not enough. The question, "At how many points of the molecular chain do the amino acid sequences of both species differ?" was framed too simply. The likelihood of an accidental mutation varies from one part of the molecule to another. To simplify matters one could say that the "stability" of the molecular structure is not equal at all points. We also have to take into consideration the possibility that mutations may have occurred repeatedly at certain places in the chain, where an amino acid was exchanged more than once. In this case a difference now recorded as single would have to be counted twice or even three times in figuring out elapsed time.

All these factors and influences make it more difficult to calculate the time span required for the exchange of one or another amino acid in the molecule. Scientists go to great trouble to correct their figures for all known and potential influences. Still no one with any experience here will deny that the results have to be taken with a grain of salt. But we can be sure that they are broadly reliable, even if in one or another case the computer's calculations may be off by as many as 20 or 30 million years.

With this reservation in mind, let me cite a comparison of the amino acid sequence of cytochrome C in humans and in chickens: it shows a period of just 300 million years during which the enzyme must have been isolated within the ancestral line of both species, without any mutual exchange, and must have been copied generation after generation. More briefly and simply put, all this means is that at the beginning of this period, around 300 million years ago, a creature must have existed that should be regarded as the progenitor both of the line that later led to the reptiles, long after that to the primates and eventually to us, as well as of the line that evolved into the modern chicken.

Studies have further revealed that our amphibious primal ancestor split off genetically from the fishes around 500 million years ago, and that the ancestors of the vertebrates and the insects must have achieved an independent genetic status roughly 750 million years ago. Far more than a billion years, probably 1.5 billion, have passed since the emergence of the common ancestor (a primeval cell), which established the connection—still present today—between us and the wheat in our fields.[20]

The reservations alluded to above affect only the absolute numbers, which might differ from the actual dates by the previously mentioned margin of error. But this does not in the least weaken the principle of the argument. In any event this method of comparing enzymes furnishes us with evidence of *the splitting up of an originally undivided evolutionary "tree," and beyond that of the chronology followed by the ancestors of modern-day species in branching off, one after the other, from the main trunk.*

The second point in particular offers us a new and really conclusive kind of proof. For if we use this chronology to work out a family tree, this turns out to be identical to those reconstructed by the paleontologists a long time ago on the basis of completely different evidence, namely from the fossils they unearthed.

Identical conclusions reached by wholly different approaches —what more convincing proof could there be for the reality of phylogenesis? On the one hand the spatial distribution of petrified remnants of extinct ancestors of modern organisms, deposited in the sediment of the earth's crust in a way that corresponds perfectly to their place in evolutionary history. On the other hand the comparison of various copies of a very old "fossil" molecule, leading by mathematical analysis to exactly the same chronology for the course of evolution. Is it possible to go on doubting?[21]

Of course, anyone who wants to put on the blinkers of prejudice will never be made to see reason by means of this argument. His last resort will be to claim that God in his inscrutable wisdom decreed that the enzyme patterns observed in the different species should display variations—variations designed so that by comparing them people would get the (erroneous!) impression that the species were interrelated and traceable back one after

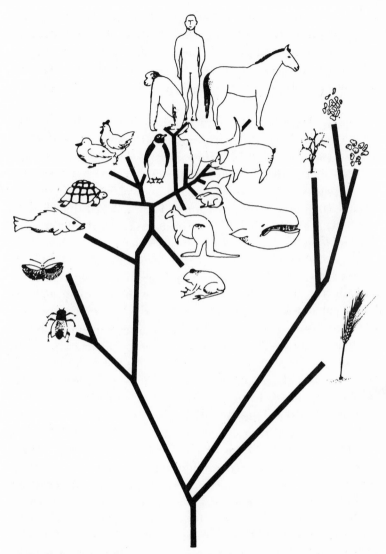

This evolutionary family tree was derived from a comparative study of the molecular structure of cytochrome C in the species shown. It is identical in all particulars to the genealogy reconstructed by means of fossil discoveries. (From an article by Margaret O. Dayhoff in *Scientific American*, July 1969, p. 86.)

another to a common ancestral line. There's no real way to challenge this statement. Perhaps, though, we may ask what's to be made of a position held by people who consider themselves religious but who prefer foisting such consistently deceptive behavior on the divine creator to correcting their own preconceived opinions.

Arguments of this sort attracted the notice of the young Immanuel Kant (1724–1804), who answered them in a way that's apt for this case as well. In his *General Natural History and Theory of the Heavens* he also deals briefly with the critics who used "theological" arguments to oppose the attempt to explain the origin of the solar system by the laws of nature. They attacked this effort as inadmissible and futile besides, because as part of God's creation the solar system was beyond natural explanation.

Thus, two hundred years ago there was something like a "religiously motivated" resistance to the possibility of explaining the facts of astronomy in scientific terms. In those days quite a few people looked on this scientific endeavor as an assault upon their world picture, a danger to their faith. For us, contemporaries of the Apollo program and the exploration of planetary space by means of space probes, such fears seem like ancient history. I for one have never heard an objection to space exploration, not even from those who are now refusing to recognize the theory of evolution for the same old-fashioned motives.

Perhaps this historical reminiscence could help us to realize that what's at stake in this biological controversy is not actually faith but only habitual ways of thinking. No religion on earth was substantially affected by the discovery that the solar system could be comprehensively described in physical terms. No one has any occasion or sees any reason for ceasing to view our planetary system as a part of God's creation merely because we've succeeded in understanding its behavior on the basis of nature's laws.

In the face of all the progress in the biological sciences, why should we repeat the same mistake that Kant's adversaries fell into? History tells us that the frontier at which (superficially) religious resistance to science begins and which since Kant's time has shifted from astronomy to biology has a tendency to wander. Isn't this an obvious indication of what's happening here? The real issue is not what aspect of scientific progress might do essential damage to the message of religion (a highly dubious notion anyway), but why there is psychological resistance to science's demand that we revise ways of thinking previously considered valid.

But now to the answer Kant gave to his critics. I shall quote it at some length, because it hits the nail on the head:

Finally, if God had directly provided the planets with projectile force and had set their orbits, then presumably they would not display the imperfections and discrepancies that we meet in everything which nature produces. Had it been appropriate for them to rotate on a plane, then presumably he would have placed their orbits precisely on a plane. Had it been right for them to approach a circular movement, then we can believe that their path would have been a perfect circle. And there is no way of telling why there should have to be deviations from absolute precision in the case of something supposed to be an immediate result of divine artistry.[22]

Kant is referring to the "minimal deviations" observable in the regular structure of the solar system, to the otherwise negligible differences between the orbital planes of the individual planets, and to the deflection of these orbits themselves from the ideal circular shape. And what he says about the meaning of those deviations is completely, unreservedly applicable to the "minimal deviations" in the structure of the cytochrome C molecule.

This is true in more ways than one. Kant too felt it was nonsensical to assume that God could, so to speak, leave these little mistakes behind after a direct manifestation of "divine artistry." And second, Kant took these irregularities as evidence pointing the way toward conclusions concerning the origin of the solar system—in just the same way that biologists have viewed the "irregularities" in the structure of cytochrome C.

What is more, I suspect that the real, unconfessed motive of people unwilling to admit that life on earth evolved under the influence of the same natural laws governing all cosmic processes must lie hidden behind their (unconsciously) fabricated arguments. I am afraid that all too many believers are flirting with the idea that they can catch God in the act of creation—that God who so stubbornly eludes our senses—by invoking a fundamentally inexplicable phenomenon which transcends the rational intelligibility of nature's laws. It strikes me that the often emotional resistance to the rational explanation of a natural phenomenon formerly thought to be unexplainable must also be understood as the refusal to relinquish an apparent "proof of God." But the hope of taking God into custody, as it were, of getting palpable assurance that he really exists, is now doomed in advance. In the area of scientific research persistent clinging to this hope leads to the painful situation where you must constantly retreat from one line of defense to the next. And anyone who does so will have to learn from the theologians that this supposed demonstrability of a Creator contradicts the image of God presented in all the great religions of humankind.

Miracula non sunt multiplicanda was a tried and tested maxim of classical Scholasticism. Freely translated, don't look for miracles when a natural explanation is ready at hand. Or, as Martin Luther put it, "We ought not to seek for God where we wish to find him, but only where he deigns to reveal himself to us."[23]

6. The Origins of Life

WE CAN no longer doubt that biological evolution is a fact, or that the species presently found on earth have *not* been there from the start, continually existing, never changing, since the beginning of the story of terrestrial life. All these organisms are the (preliminary) result of a long history of development, still going on today, and no serious scientist questions this. And recognizing the historical reality of phylogenesis opens our eyes to a still more comprehensive and meaningful context—its connection with the prior stage of cosmic evolution.

For it is just not true—as an exclusive preoccupation with biological development might suggest—that the evolutionary process only began with the coming of life. Nor, still more importantly, is it true that cosmic and biological evolution have nothing in common except the word evolution, that they are two essentially different processes. On the contrary, one of the most exciting discoveries of modern science is that drawing on the gigantic number of its particular findings, like countless mosaic chips, it is piecing together an increasingly clear picture of a world where everything dovetails with everything else, the largest with the smallest, the farthest with the nearest, and even the dead with the living—in other words, a homogeneous, self-contained world.

Actually, with the aid of hindsight we can call this a truism, but even so it comes as a surprise. And many people react to it at first with skepticism if not inner resistance, because the picture, whose outlines we can now make out, contradicts (precisely on account of this self-sufficiency) the ways we've thought about the world for hundreds of years.

For a long time we had gotten used to dividing the world up into different zones, to breaking down its "contents" into categories, which, we believed, had nothing to do with one another. On the one hand the vast expanse of the universe; on the other, a tiny speck lost in this immensity and unrelated to it, our earth, only tolerated, as it were, because of its insignificance. Even Jacques Monod, the French biologist and Nobel prize winner, felt obliged to stress how desolate and insurmountable the distance between these imagined opposites was.[24]

Putting it another way, on the one side was man and on the other, supposedly separated by an absolutely unbridgeable chasm, the rest of animated nature. We even misunderstood our earthly environment (as is steadily becoming more apparent) by viewing it for all too long as a backdrop, at once unlimited and infinitely disposable, for human activity. We mistakenly thought that we had somehow been thrust "from outside" into the world and hence fancied ourselves exempt from its laws.

The course of scientific investigation has shown all this to be a misreading of our place in the universe. The world does not consist of "zones" governed by different sorts of laws. It likewise has no "contents." Everything that exists is "the world," part of the one and only self-contained reality and connected within it to every other part of the whole.

The boundaries we think we see everywhere don't belong to the world itself. They are nothing more than projections of our imagination, something like a grid that our brain slips over the exterior world to help us get an overview of the teeming spectacle before us. (After all, the grid on a topographical map, which serves the same purpose, doesn't reflect any actual features of the landscape.)

The fact that scientific research is so compartmentalized does not reflect any compartments in nature but only our inability to see the universe whole and study it as such. The disproportion between the complexity of the world and our limited brains leaves us no other option but to isolate individual aspects in the totality of phenomena. But we get a double dose of our incompetence if we let ourselves be misled into thinking that just because science keeps splitting up into more and more specialized disciplines, there must be a corresponding fragmentation in nature.

This holds for all such "boundary lines" that we seem to find in the natural world. For many critics, especially those in the humanities, the mere thought of crossing these lines is a mortal sin. Of course, there *are* some kinds of boundary infringements, above all with regard to method, that can't be allowed. Still, we have to object to these critics, up till now the most important insights we've gained about nature have been in effect successful violations of boundary lines.

More than a hundred years ago the boundary between the laws of classical mechanics and those governing the theory of gases was considered inviolable. When it was lifted, however, it provided the impetus for the development of modern thermodynamics, ultimately leading to the concept of entropy[25] and a deeper understanding of all temporal processes in nature.

Up until 1828 everyone took it for granted that organic and inorganic chemistry were separated by an uncrossable barrier. Everyone, including scientists, assumed that organic compounds, unlike inorganic substances, could never be "artificially" produced in the laboratory, only by means of a biological synthesis in a living organism. But then Friedrich Wöhler proved that urea could be synthesized in the laboratory, and in so doing laid the foundation for a new and immense field of research in organic chemistry.

It has always been the same, from Newton's discovery of the connection between the laws of gravity as observed on earth and the laws of planetary motion, to Einstein's theory of relativity, which puts to flight our innate notion of the unbridgeable boundary between time and space. One could write the history of natural science by describing all the triumphs over boundaries (erroneously thought to be real) between natural phenomena that strike the human imagination as fundamentally different.

Needless to say, this history is by no means finished. Our psychic constitution requires us to undertake essentially the same task all over again whenever a new boundary line comes into view. And however willing we may be in advance to call it an illusion, the enormous job of proving it so remains to be done. One of humanity's greatest intellectual achievements is recognizing reality "for what it is" (a goal that by definition—to anticipate a later point—can never be completely achieved). Science is nothing but the attempt to go as far as possible along this road, to overcome our instinctive modes of looking at the world, to abstract from them in the true sense of the word, to get beyond subjective appearances and to lay bare a little piece of objective, "real" nature.

For some time the decisive frontier in contemporary biological research has been the one between animate and inanimate nature. Anyone who calls this boundary into question, that is, anyone who believes in the possibility of a "natural" and hence basically intelligible transition from inanimate to living material structures, immediately runs into objections from the so-called vitalists.

"Vitalism" is the conviction that all natural processes and in particular the emergence of the first life forms are essentially beyond the power of science to account for. Putting it positively, vitalism holds that life is fundamentally different from all other natural processes, especially physical and chemical ones, and must be viewed as the expression of a special "life force," which cannot be grasped scientifically or in any other way.

This notion, likewise derived from a "natural boundary," must sound almost depressingly familiar to anyone with a sense of the history of science. But vitalists learn nothing from historical experience. For more than a century they have been on the defensive, steadily retreating, but this hasn't made a dent in their obstinacy. Experience shows that a convinced vitalist gives ground by the millimeter and only when the facts are staring him irrefutably in the face.

Now no biologist will deny that there are gaps in the proof that life came into existence on the earth in a natural way, that is, in a spontaneous transition, controlled by natural laws, from inert to animated matter. In the mosaic that scientists have patiently begun to assemble there are still quite a few important chips missing—a fact that gives the vitalists no end of satisfaction. But the outlines of the picture are already well defined, and we can safely predict another vitalist retreat. Let's take a look at the data which would lead an unbiased observer to expect that sooner or later the apparently monumental barrier between animate and inanimate nature will be shown to be illusory.

We can begin with the oft told story of the Miller experiment. In 1953 the young American chemist Stanley Miller, a student of the renowned Nobel prize winner Harold Urey, enclosed in a glass container the simple inorganic molecules which, as he had learned from his teacher, should have been abundantly present in earth's primeval atmosphere: carbon dioxide, methane, ammonia, and molecular hydrogen. He kept the solution in a state of agitation for several days, while causing it to evaporate and condense in constant succession and at the same time subjecting it to electrical discharges as an external source of energy (to simulate the violent thunderstorms that must have broken out in the primordial atmosphere).

Every high-school student who has taken even a halfway decent course in biology knows what happened next. Miller's incredibly simple attempt at imitating circumstances present on the surface of the (as yet lifeless) earth resulted in the spontaneous emergence of some of the most important building blocks of life, amino acids in particular. Up till this point the production of such "biopolymers" under nonbiological conditions had been viewed as something that would be especially hard to explain, if not impossible to begin with. Then Miller demonstrated how very wrong such views were.

At the time reports of this experiment understandably struck people as sensational. Today, less than three decades later, the Miller experiment has become a standard procedure in some of the better biology courses given to high-school seniors. And the theory illustrated by Miller's experiment has likewise gone in a short time from a sensation to a self-evident, everyday affair.

Once again we can hardly doubt that only our prejudices made us assume for so long that the emergence of these and other basic elements of life was an "absolute mystery." It was prejudice that caused even scientists to flirt with the intellectual resignation of vitalism and to consider seriously the idea that a nonbiological source for the molecular building blocks essential to living organisms might in principle defy our powers of explanation.

Nowadays all these doubts have shattered. Miller and his successors have shown that the real situation is totally different from what our prejudices would have us believe. The spontaneous emergence of the biopolymers (building blocks of life) is anything but mysterious and inexplicable. On the contrary, it's "the most natural thing in the world."

It's obvious that on account of the peculiarities of its atomic structure matter is inclined, "prefers," to unite, at every possible opportunity, into compounds we have come to know as the molecules of life. Their spontaneous coming into existence is not only no longer mysterious; given the influence of the natural laws governing the behavior of matter, the emergence of life seems to have been altogether necessary and inevitable. It should be immediately noted that this in no way makes the whole process less wonderful. The mystery still remains, for how would we ever manage to explain *why* matter is so constituted that it necessarily produces the chemical conditions required for life?

This reading of Miller's experiment and its countless later variations (which have furnished practical proof that all needed biopolymers can be generated from nonliving matter) has been confirmed by some illuminating discoveries in a very different scientific discipline, namely astrophysics. After a series of accidental finds, around 1970 radio astronomers began systematically looking for chemical compounds in outer space, to be more precise in the gigantic interstellar dustclouds. This quest was overwhelmingly successful with regard to both the number and kinds of molecules discovered. A recent survey lists no fewer than 30 organic compounds whose presence has been established in the space (once considered empty) between the stars of the Milky

Way. Taking a closer look at the list we note with amazement that it's made up almost entirely of molecules familiar to biochemists as preliminary constituents of the building blocks of life.

Formic acid and monoaminomethane molecules, for example, were found in a filmy cloud of gas several thousand light-years away. These two molecules combine to produce glycine, the amino acid that occurs most frequently, biologically speaking, as an element in proteins. A similar situation obtains with highly specific (on the chemical level) sugar molecules that are components of the complex molecule involved in heredity, ribonucleic acid (RNA), which is present in all living organisms.

The British astronomer Fred Hoyle has correctly pointed out that under these circumstance we can assume that far more complicated molecules (up to complete amino acids and ribonucleic acid itself) spontaneously come into being in cosmic gas clouds. For physical reasons, however, such molecules cannot be detected with the technology currently available to radio astronomy.[26]

However thinly these molecules may be distributed throughout interstellar space, a cosmic cloud stretching for several light-years contains vast quantities of them. Some years ago I suggested the possibility that the molecular components indispensable to the emergence of life on earth might not have developed on the surface of the primeval earth—as almost everyone thought back then. Perhaps, I conjectured, they were brought to our planet from every quarter of outer space, in the form of cosmic dust or transported by meteorites. When a planet, thanks to gravitation, picks up molecules from the space around it like a sort of seed crystal, then the yield must be enormous.[27]

Fred Hoyle has recently gone far beyond this possibility and formulated the hypothesis that even the first primordial organisms capable of self-replication were not formed on earth but, perhaps, in the heads of comets. After these burst asunder (which was inevitable, sooner or later, owing to the cosmic forces present in regions close to the sun) the organisms could have survived the plunge through the atmosphere and the landing on the earth's surface by virtue of being in the interior of chondrites (meteoric stones). This hypothesis has been strengthened by the work of a team of American researchers. Not long ago they came out with data that make it appear likely that in the head of a comet "over a biologically sufficient time span" environmental conditions prevail that favor the emergence and even the evolutionary development of such seeds of life.[28]

Since in all probability at least 100 billion comets revolve around the sun, the number of openings here for "biological experiments" is so great that these days the experts are discussing Hoyle's theory very seriously. It could be that we have to imagine the first appearance of life on earth as springing up from seeds sown by the cosmos.

Once we begin to consider the matter in this light, the scales fall from our eyes. We're struck, for example, by the fact that chemical analysis of stone meteorites has produced a truly breathtaking correlation: the frequency distribution of the amino acids contained in these meteorites is practically identical to the distribution that shows up spontaneously in all the experiments à la Miller. (Quite apart from that it's exciting enough that these investigations of the cosmos have turned up a large number of amino acids, and flawless ones at that.)

A few examples: chemical analysis of the so-called Murchison chondrite (named after the site in Australia where it was found) identified no fewer than 17 amino acids. Ten of these do not occur in living creatures. No less than a third of the entire amino acid content of this stone, which literally dropped out of the sky, was allotted to glycine. In second place was alanine, a protein building block, and in third and fourth place aspartic acid and valine.

In all chondrites studied thus far where amino acids have turned up, they have basically followed this order. (Sometimes first and second place are reversed.) That fact is remarkable enough by itself, because all these meteorites come from different regions of space, depending upon the point in time when the earth picked them up during their journey with the sun.

Once again aren't we forced to conclude that all throughout the universe the same elements everywhere combine to form the same compounds, and that certain compounds (glycine or alanine, for instance) are evidently formed with greater ease than others?[29]

This interpretation seems irrefutable, especially when we consider that the sequence of amino acids observed in the meteorites matches the relative abundance of the same protein elements generated in Miller's experiment (and all its variants). The chain of proof becomes complete when we add the fact that this order of priority is identical to that found by biochemists in the cells of all living creatures on earth. Glycine is the most frequently used protein building block in all plants, animals, and in our own bodies. Then comes alanine, not only in meteorites but also in the spontaneous syntheses carried out in the laboratory and in every cell of living matter on earth. The same is true of the next most frequent elements, aspartic acid and valine.

The message in these results is obvious. Once more a problem that was presumably insoluble by definition has been shown to be no problem at all, a pure phantasm.

The issue here was how to explain the apparently "absolutely incomprehensible" fact that at the moment when life first arose on earth precisely those 20 amino acids which nature needed to construct living organisms were ready and waiting. "There are hundreds and hundreds of different amino acids," objectors tell biologists. "How can you ever explain why, at a time when the earth was still dead, just the "right" 20 should be gathered together, and in just the right quantitative proproportions?"

Once again the vitalists appear to have an airtight case that reduces to absurdity science's working hypothesis that all natural phenomena can be explained by natural laws. In point of fact, though, how could this question ever be answered in natural terms, without appealing to supernatural causes? It turns out that the way the question is posed is, quite simply, false—and unanswerable. It sounds impressive, but it collapses as an argument as soon one sees that in its present form it's of no interest at all, that it has nothing to do with the reality we're trying to explain and understand.

Our tacit assumptions were all wrong. It was not the case that life on earth was dependent on only 20 quite specific amino acids (and, in addition, countless other molecular elements) from among hundreds of possible ones. The way the "molecular palette" of the cosmos matches the earth's biological one forces us to look at the situation from a totally different perspective.

At the decisive moment when the earth's history began there was, quite obviously, no specific *need* for future, as yet nonexistent, organisms; there was only the necessary material. In other words, we have to assume that the reason why the 20 amino acids are found today in the cells of all living organisms is not that there was no other way for life to get started. Rather, all the data we have seen thus far suggest that earthly organisms were constructed with the help of these amino acids because for reasons given they were present in great abundance and thus available as building blocks.

The atomic structure of the 92 elements found in the universe is different in each case. Some of them have only a slight propensity, or none at all, to combine with other elements. From ancient times these have been known as "noble" elements, because they tend not to "mingle" with others, for example the noble metals gold and silver, or the noble gases helium or argon.

Other elements have quite a different disposition. Hydrogen and oxygen, for instance, combine, at every opportunity as it were, to form molecular compounds. There are also marked affinities between certain elements, leading them to react by preference with each other even when other elements are present.

The modern physical chemist can plausibly explain these various constitutional "inclinations" by means of specific features observed on the surface of atoms of different elements. Putting it simply, the helium atom, for example, has a smooth, fully closed surface, making it altogether unfit to pick up a second atom. By contrast the surface of an oxygen atom displays optimal "fits" for a whole series of atoms. Analogously, the affinities between certain elements can be explained by certain structural congruences in their external electron shells, as has long since been demonstrated by experiment.

It is clear that these peculiar features must exert a definite control over the reactions caused by combining various elements. "Noble" elements will not be affected by contact, while all others will, depending on the circumstances, combine with appropriate partners to form molecules.

"Depending on the circumstances" means that the temperature, pressure, and concentration of the mixture also help determine the course of events. This process (which has been extremely simplified here) is further complicated by the fact that the affinities and peculiar reaction patterns of molecules (which are continually being formed from only a few atoms, wherever possible as few as two or three) differ from those of the elements entering into them and undergo unforeseeable changes, depending upon their structure.

All this results in a very complicated process, at the end of which—when conditions are right—there will be an abundance of the most varied combinations, some of them already relatively complicated. The limits of the "right conditions" should not be thought of as overly narrow, as is shown by the productivity of laboratory arrangements as simple as Stanley Miller's, together with the high incidence of such molecules in free space. The results of Miller's experiment as well as cosmic synthesis in space are (broadly speaking) "always the same." This suggests that this course of events is by and large determined by the peculiarities of the structural elements of the cosmos and by the similarity of external physical conditions throughout the universe.

The key to the sameness of results is the spontaneous and inevitable (in view of qualities of matter and the laws of nature) emergence of biopolymers, the building blocks of life. This is the conclusion reached by contemporary science, and it is genuinely exciting. It helps us to see that the transition from cosmic to biological evolution is a seamless one. They are, to repeat, one and the same process. We only thought we had to look at them separately, because we had long ago (to get a better overview) given them separate names. But this was centuries before we got the chance to discover the connection between them.

We have to emphasize strongly that this stirring discovery, which is undoubtedly of the greatest philosophical significance, would never have been made if scientists had yielded to the line taken by the vitalists. For the latter had maintained right from the outset that it was "impossible" for life to arise from natural causes, that this could only be imagined as taking place with the aid of supernatural factors.

The list of unsolved problems (which the vitalists always declare insoluble) and of apparently insurmountable difficulties was in fact intimidatingly long. One obvious option was giving up. Given the seeming hopelessness of the task this would have been understandable, even pardonable. But scientists would not give up. They stubbornly clung to the intention underlying all their work—to find out how far they could get in their effort to understand nature without recourse to miracles.

And here too their stubborn adherence to this principle paid off. Their reward was an insight that taught us to see our role in nature, our place in the cosmos, with fresh eyes. Now for the first time there were data pointing to a concrete link between events in the universe and life on earth. Some ten years ago the French Nobel prize winner Jacques Monod spoke with suggestive pathos of "the meaninglessness of human existence"[30] against the background of an alien cosmic reality. For the first time doubts were cast on that dramatic pronouncement. Now it appears that this cosmos, supposedly so alien and hostile to life, might just be our cradle.

These insights have tremendous power; they broaden our picture of the world in surprising ways and deepen our self-understanding. But we would have to do without all this—and much more—if the creationists and vitalists had managed to have things their way. A hundred or more years ago we would have shrugged our shoulders and settled for the answer that there was a "miracle" here. There would have been no reason to seek any further, to continue the tedious task of research.

What a wretched hypothesis the vitalists and creationists have been trying to thrust upon us all these years. Upon closer inspection we can see that it isn't even a hypothesis, because vitalism has no concrete, palpable content. It is reducible to the simple bias which pretends that all phenomena unexplained thus far must on principle be unexplainable, and which derives from this the "proof" for the efficacy of supernatural forces.

New scientific discoveries keep proving the vitalists wrong, but they aren't troubled in the least. Silent and unblushing they merely take a step backwards, onto still unexplored terrain, whence they go on reciting their familiar line without a moment's hesitation. It's hardly comprehensible at first glance why so many people still think vitalism a plausible philosophy, and why, even though this monotonous credo has been steadily refuted down to the last detail over the past hundred years and more, it remains popular, at least among lay people.

The best explanation for this probably lies in the fact that vitalists pretend to be taking a *religious* position. They claim to be defending the religious view of the world against the advance of "materialistic" science. To the degree that they succeed in spreading this impression around, they understandably strike a responsive chord.

But the claim that any circumstance, once explained by science, ceases to be a possible object of wonder, including religious wonder, is simply nonsensical. Had vitalism carried the day, it would have deprived us of all the scientific findings mentioned above. But, above and beyond that, vitalism undeniably gives rise to an extremely dubious sort of theology, for religion speaks to the real man or woman in the real world. Anyone who falsifies our picture of humanity or the world by suppressing knowledge that would allow us to see their reality in sharper focus is *ipso facto* also falsifying the contents of religious statements that deal with humanity and this world.

Hence, despite the sympathetic support it has all too long received from certain church groups, vitalism has been a disaster, not only for science but for theology too. This claim, of course, has to be substantiated in detail—and will be in Chapter 10.

But, on the other hand, none of this should be taken to mean that scientists consider the world and humans perfectly intelligible, nor that it's only a matter of time before human ingenuity solves all of nature's puzzles. Back around the turn of the century, in the so-called classical era of modern science, there may have been a few individuals who toyed with this possibility, but it's now passé.

The modern—which means evolution-minded—thinker can only judge as downright absurd the notion that after a development that has lasted for eons our brain should have managed, precisely at this moment, to reach a degree of perfection allowing it to comprehend the entire universe in its objective reality. (And the idea that we might be able to achieve this goal sometime in the historical future is every bit as absurd.) We must, therefore, proceed on the assumption that scientific progress will not go on indefinitely at the same rate of speed that has marked the previous centuries. Sooner or later we shall surely come upon frontiers where nature *will* once and for all defy our powers of explanation. In many areas, especially in physics, we seem to be getting the first indications that we're already approaching such a frontier.

But this experience should not be read as evidence for the abrogation of the laws of nature by "supernatural factors." It's something that was altogether to be expected: the consequence of our brain's inadequacy to cope with the greatness of nature. And, apart from the limitations of our intellectual capacity, we ought to feel secure in crediting nature with the ability to function naturally. Vitalists don't take this possibility into account—yet another error on their part. Vitalism is also bad philosophy.

Except for the possible case of physics just mentioned, there are no definitive boundaries in sight to our quest for knowledge so far. There is undoubtedly still an enormous amount of room for the unfolding of our scientific knowledge of nature. Limited as our brain may be, there are still innumerable questions, problems, and puzzles for it to attend to, all of which are worth the effort, and whose solution, given enough patience and perserverance, is only a matter of time.[31]

So there's really no need to lay any special stress on the fact that nowadays we're still a long way from fully understanding the origins of life on earth in all its complex stages. No one ever denied it. Vitalists, who always think they're "proving" something when they point, with a barely contained sense of triumph, to the incontestable abundance of questions still open and unresolved, are thus going into raptures over the flattest of truisms.

As of today this much can be said: all—that is, every single one, without exception—all indications suggest that for the foreseeable future the stubborn persistance of scientists will be rewarded in the area of the origins of life too. As in the case of the biopolymers, so with most other problems that need to be solved in this connection; for some years now the outlines of a possible scientific solution have been coming to light. This book cannot and does not intend to handle these other problems with anything resembling thoroughness. Readers interested in the current state of research in this field are advised to consult some of numerous books published on the subject in recent years.[32]

But, to conclude this chapter, I want to say something about two questions in particular. The first concerns a basic problem that until a very short time ago seemed almost insoluble, but that Nobel prize winner Manfred Eigen appears to have cracked. The second concerns an incidental pseudo-problem. But both are important in the context of our discussion because lately certain people have advanced them, with much fanfare, as "proof" that life could not have originated from a natural source and have thereby caused much confusion among nonscientists.

The first problem relates to the origin of genetic coding in the various amino acids. Putting aside the details, we need only recall that the molecule for heredity, ribonucleic acid (RNA), contains a sort of text or cipher or code that specifies how the amino acids are linked together and hence how they can be assembled to form specific proteins such as enzymes. The "letters" of this text consist of certain chemical combinations, called bases. Nature makes use of only four different bases to draw up its blueprint. In RNA these are adenine, cytosine, guanine, and uracil. It's the same in DNA (deoxyribonucleic acid), with the sole exception that thymine replaces uracil. For simplicity's sake biochemists abbreviate the names of these bases as A, C, G, and U or, in the case of DNA, T.

Any three of these bases (known as a triplet) "mean" a specific amino acid. For example, if the bases cytosine, uracil, and adenine make a sequence in an RNA chain, this is equivalent to an order to add on the amino acid leucine to this section of the blueprint. In the same way the triplet of guanine, uracil, and guanine "code for" the amino acid valine, while uracil, cytosine, and guanine code for another amino acid, serine. In the abbreviated language of biology, this means that CUA codes for leucine, GUG for valine, and UCG for serine.

All this has been known for many years and is in every college biology textbook. But from the very beginning this discovery has given grief to all the theorists who devoted themselves to the difficult task of explaining the origins of life in scientific terms. Their problem was how to account for the way the different base triplets had come to have their respective meanings. Why, that is, does the sequence UCG code precisely for serine (and not some other amino acid)? Why does GUG "mean" valine? Why does the base sequence CUA represent the "information" that determines the formation of leucine? How should any triplet at all have acquired a "meaning," in the sense of providing such information?

The relationship between the individual triplet and a specific amino acid seemed to resemble the one in Morse code between a given sequence of dots and dashes and a specific letter of the ordinary alphabet. In Morse code a single dot means "e," two dashes mean "m," while the series comprising a dot, a dash, and two more dots codes the letter "l."

This comparison throws our problem into sharp relief, because a dot in the Morse alphabet (like all the other symbols in it) came to mean what it does only through an arbitrary decision, a prior agreement among the people who wished to use this artificial alphabet for specific purposes (such as radiotelegraphy). Between the Morse code symbols and the letters of our everyday alphabet there exists, to use a slightly highfalutin term, a semantic connection. The information contained in a given Morse sign can only be thought of as the product of arbitrary convention.

The reader can see where this argument is headed. Wouldn't things have to be exactly the same with the molecular "signs" of the genetic code? Doesn't the meaning of a given base triplet have a semantic character? But in that case how could the laws of nature explain the origin of the information contained in this meaning? Here too wouldn't there have to be an "agreement," an "arbitrary determination" before such information could emerge? And, if so, who could have actually done this "agreeing" and "determining"?

It's not surprising that the vitalists seized on this point immediately. Two of them in particular, A. Ernest Wilder Smith and the biologist Wolfgang Kuhn, have been harping for years on the argument that the coding process cannot have been nature's work. Information, they insist, is inconceivable without some sort of prior agreement between the partners exchanging it. (Wilder Smith especially likes to talk about the need for "know-how." He never defines precisely what the term is supposed to mean in this context, but it sounds impressive, and one is free to interpret it any way one wants.) They argue further that since the base triplets of DNA and RNA, as biologists themselves have ascertained, contain a unit of concrete information, this requirement must hold for them too. Thus, reversing the view of most scientists, they conclude that the meaning of the genetic code in the biological process of protein synthesis cannot have come into existence spontaneously, by purely natural causes, but must have been arranged, by God, Wilder Smith would say. "You not only can believe in God, under these circumstances you must believe in God," to quote (roughly) the characteristic ending of one of his recent lectures on the subject.

But, alas, God is not so easily gotten hold of, even with the help of the genetic code, as every theologian will admit. In other words, Smith and Kuhn's position, though decked out in phraseology sure to impress the layman, is untenable. Their publications, if they attract any attention at all in scientific circles, can't stand up to serious criticism for an instant and simply leave readers shaking their heads.

Both authors are either unaware of or have failed to grasp the scientific concept of information (and information theory). Information in the scientific sense is *not* the same thing as a piece of information in the everyday sense. The latter kind is found only when the meaning of the signs bearing the information is known to the one who perceives them. Thus a sentence in Chinese will most likely convey no comprehensible information to a Westerner who hears it. But to apply this "naive" sense of information to the scientific term (as in "genetic information") is an error that a layperson might be forgiven, but not a biologist.[33]

Information, as defined by information theory, is the measurable deviation of signal distribution from the statistical average. It has nothing whatsoever to do with the "content" of what we normally call information.

I admit that such an abstract, purely mathematical-statistical use of the term is not easily understood. It's also quite impossible (but neither is it necessary) to explain it here in a few lines. Anyone who wishes to learn more about it will have to refer to one of several introductory guides to the subject.[34] The name "information theory" is surely rather unfortunate, as even experts in it concede. It probably would have been more to the point to speak of a theory of the processing or transmission of signals.

A layperson is not expected to know this, but a scientist claiming competence here really must. An astronomer who thought that the Milky Way, because of its name, was a dairy product would be laughed to scorn. But Wilder Smith and Wolfgang Kuhn are caught in a similar mix-up when they attack the "official doctrine" concerning the effective information in the nuclei of cells.[35]

Base triplets and amino acids are not related to each other in the way, say, the sender and recipient of a telegram are. Their sort of "information" has no semantic character and is in no way "exchanged" by the two partners. And, again, it has no content in the usual sense, thus eliminating in advance the need for an agreement, of whatever kind, to ensure its efficacy.

What then is left of the term? A scientist might give the following (oversimplified) answer: the DNA molecule in the nucleus of a living cell contains information insofar as the sequence of base members does not correspond to a purely random distribution. The base sequence of DNA deviates from the statistically average distribution in a mathematically measurable way.

Still more simply: the distribution of bases within a string of DNA is unlike the "optic noise" that appears on a TV screen when the set is on but the station is no longer transmitting. Instead it resembles a recognizable pattern. It should be noted in passing that this nonrandom distribution of the bases of a DNA string appears not only in the living cell but also quite spontaneously, under the influence of natural forces, when a strand of DNA artificially synthesized in the laboratory is left to grow by itself.

Experiments conducted in the Manfred Eigen Institute have demonstrated that this is simply the result of various "affinities" displayed by different links in the molecular chain. Certain links or members, when present at a particular point of the molecule, enhance its stability and hence its chances of survival in the test tube. They are thus superior to other arrangements, where the neighboring members of the chain are less chemically compatible and so disintegrate more quickly in their watery environment.

Here we have genuine evolution taking place, in a test tube, on a level that is still purely chemical and prebiological: accidentally emerging molecular variations are selected by the environment on the strength of their long-term viability. This experimental discovery obviously points up the fact that evolution is not, as we all too often thoughtlessly presume, a specifically and exclusively biological process. Here too we recognize the effective power of one and the same principle, which overflows the bounds that we project into nature.

In this way DNA chains with quite specific sequential patterns are formed both in the test tube and, of course, in nature at predictable rates of frequency. Scientists call this sort of pattern "information," and so these molecules "contain" information, which (it's important to add) has arisen spontaneously.

This information is, in the ordinary sense, absolutely meaningless. What has happened, simply speaking, is that a molecule has come into existence that stands apart from the average because of certain "striking features" of its makeup. These structural peculiarities of the DNA molecule make it like a key for which, as yet, no lock exists. Thus the problem facing molecular biologists was to discover how nature managed to come up with a usable lock for this key.

This, and no more, is what is meant when biologists say that a base triplet contains information for attaching a specific amino acid, and when they ask how the triplet could have acquired the distinctive feature of coding for just this amino acid rather than another. The whole process obviously has nothing to do with what we usually call information. Nothing is exchanged between the triplet and the amino acid, there is no message, and therefore no need at all of a prior semantic convention.

Hence Wilder Smith and his associates are merely revealing their ignorance when they accuse Manfred Eigen of "nonsensical" claims when he uses the term "information" in the scientific sense just explained.[36] Actually, it's an open question whether what we have here is simple ignorance or something worse, namely deliberate deception on behalf of ideological prejudice. Whether or not the vitalists ever succeed in understanding or properly explaining information theory has in this case long ceased to be of any significance.

Manfred Eigen and his co-workers have discovered the first indications of how the information link between base triplets and amino acids is to be understood. Their findings are as simple as they are enlightening. It turns out, to begin with, that the triplet

GGC (i.e., the sequence of guanine, guanine, and cytosine) is the most chemically stable of all conceivable base combinations and, at the same time, the triplet whose molecular structure least resembles that of all other combinations (thus rendering it the most specific of all available keys, the one least likely to be confused with another).

But GGC codes for—of all things—glycine, which happens to be the most frequently encountered amino acid. This can't be an accident, as proved by the fact that the same rule holds for the informational connection between the triplet GCC and the amino acid alanine: after GGC GCC is the second most stable and most specific triplet, and alanine, as already noted, is the second most frequent amino acid in nature. Further research has turned up the same link between aspartic acid and valine and their respective coordinate triplets.

So at this point it looks as if evolution, in its opportunistic way, simply seized on the frequency distributions of both sorts of building blocks to establish the coding connection we now find between them. Once again a problem that just recently seemed so mysterious appears now on the verge of being solved—a marvelous solution, to be sure, but a perfectly natural one. In any event we can theoretically infer how it is that in the concurrence of ribonucleic acids and amino acids during the initial phase when life emerged the currently fixed coordinates could have come about simply as a result of probable contacts.

Whether or not the vitalists comprehend this or are willing to accept it, we see here a concrete step forward in our understanding of nature—and, in the final analysis, of the laws responsible for producing *us*—which compels the vitalists to retreat another step backwards. But they shouldn't find it difficult—they have had plenty of practice.

The second question, the pseudo-problem I mentioned earlier, can be quickly dealt with because its solution, as we have said, is completely obvious. It needs to be mentioned only because it too has been loudly heralded as evidence for the supernatural origin of life and as proof for the existence of God.

The problem consists in the "handedness" or, to use a fancier term, the chirality (from the Greek *cheir,* "hand"), of certain elementary units of life. Thus all the amino acids that occur in animate nature form left-handed helixes (like a corkscrew, only reversed), and the nucleic acids in all earthly organisms form right-handed spirals.

This state of affairs, as many vitalists see it, can't be explained in natural terms, because when both kinds of molecules are produced naturally (spontaneously) in the test tube or inanimate nature, you always get a "racemate." (That is what chemists call a compound that has left-handed and right-handed helical molecules in equal proportions.) Up to this point their argument is altogether correct.

But the vitalists go on to insist, once more, that we must assume the existence of a "planning spirit" with the necessary know-how to perform the exacting task of selecting and assembling out of the spontaneously produced compound the "levo" and "dextro" helixes indispensable to living organisms. One writer even claims that in most textbooks this problem is hushed up because the authors are only too aware that it runs counter to their "materialistic viewpoint."[37]

The reason for this "hushing up" is, however, far more innocent: the issue of chirality, allegedly such a decisive factor in the origin of life, actually poses no insoluble questions for biologists, and so need not be discussed in technical publications. In Eigen and Winkler's book, *Laws of the Game,* which is addressed to the layman, readers can find a detailed answer, which is at bottom extremely simple.[38]

Briefly it goes like this: the first systems capable of self-reproduction were certainly not yet "alive" in the usual sense and were most emphatically not cells. They were probably molecular systems like the hypercycles described by Eigen, which were already competing for the limited amounts of elementary materials need to make copies of themselves.

And in the course of this competition the successful molecular variants were the ones that as a result of random combinations of qualities turned out to be less destructible ("more long-lived") or reproduced at an above average rate, that is, either more quickly or more economically than the mass of their competitors. Evolution takes place, as we have seen, on the prebiological, molecular level.

Any systems that would have built themselves up out of the existing racemate by making indiscriminate use of both left- and right-handed (helical) amino acids (or nucleic acids) would have been at a hopeless disadvantage from the very start, because for every single stage of assembling proteins from individual amino acids they would have needed two enzymes instead of one. Enzymes, we recall, are spatially structured molecular keys. Any one enzyme, therefore, fits only one left- *or* right-handed amino acid (or nucleic acid).

So during the very first phase of development—even before enzymes came into play—they would have had a smaller chance of surviving. For a protein made up of a racemic compound of amino acids that coiled in different directions would evidently be less stable than a "pure-blooded" left- *or* right-handed helix. For this reason, among others, it is most unlikely that such racemic systems would have participated very long in the contest among molecules that marked the initial phase of life.

We can assume, then, that during this stage the struggle for survival was joined exclusively by systems whose proteins and nucleic acids formed, respectively, either all left- or all right-handed helixes. Of these competing systems only one has remained. Its protein was constructed exclusively of "levorotatory" amino acids and whose nucleic acids were composed exclusively of "dextrorotatory" elements. We know this for certain because the descendants of this one variant are the sole survivors present on earth today.

On the other hand we can no longer determine why precisely this system won out in its time and beat all the competition. We can only say that this surely happened at a very early point in time, probably before the beginning of actual biological evolution. The explanation must lie in some quality that gave this variant a considerable advantage over its competitors. This may have been due, for example, to the acquisition of an enzyme that ensured a faster rate of reproduction or enhanced the reliability of the copying process.

In such a case the privileged variety must have prevailed very quickly. Eigen gives a graphic picture of how these processes occurred in the "evolutionary games" that he describes in his previously mentioned book. In any event, despite all the difficulties that this matter undoubtedly still presents, the asymmetry of life's building blocks has long since ceased to be an insoluble problem.[39]

7. Darwinism

THERE IS no population explosion going on among robins—a fact that, strangely enough, surprises no one. Nor is anybody amazed that seagulls, crows, sparrows, and all other birds (and animals) don't rapidly multiply on such a vast scale as to make the flocks in Hitchcock's *The Birds* look harmless by comparison.

One of the basic characteristics of genius is its ability to discover questions and look for answers in places where we normal folk have been dulled by habit into taking everything for granted. The "robin problem" is an instructive example. It was one of the things that drove Charles Darwin to develop his bold theory, which has revolutionized our understanding of the world.

The issue is so simple that a schoolboy can grasp it. All one has to do is confront the facts and see their consequences. It's simple arithmetic: a pair of robins produces about 10 eggs a year. With a life expectancy in the wild of about three years the female robin would lay 30 eggs. If every one of these eggs hatched, the total number of robins would increase by fifteen-fold within only three years (more than that, actually, since the young would begin to mate and multiply even before their parents died). And for this larger robin population the same rule, of course, would hold—a 1500% increase every three years. Within a few years the sky would be black with robins. But this doesn't happen—why not?

The answer is as simple as the computation. The 30 eggs laid by the one robin don't turn into 30 new robins. Some eggs don't hatch. Some chicks fall out of the nest or freeze to death after a cloudburst. Some helpless fledglings are eaten by cats, and so on.

A familiar story. But have you ever thought what the overall rate of loss might be? The figure is distressing: out of 30 eggs, on the average, only 2 young robins survive long enough to lay the full quota of 30 eggs—2 out of 30, which is over 93% lost.

The numbers derive from the observation that for the most part the robin population remains constant over the years. This could only be the case if each pair at its death left behind exactly two descendants, no more and no less, exactly the number, there-fore, needed to make up for the parents' death. The robin popu-lation could not remain basically constant unless this were the average breeding career of a pair of robins.

Hence there is no arguing over the rate of loss. One may find the figures shocking—they don't jibe with the idyllic pictures we like to paint of nature—but there's no room for disputing them. It's also obvious that a like pattern (in most of the lower animals things are far worse) obtains, for precisely the same reasons, among all other living creatures on this planet.

All this is perfectly commonplace, and the biologists of Dar-win's time were well aware of it. But Darwin was the first to ask the decisive question whether reasons might not be found to determine *which* two offspring of a given pair would survive, given a steady population. Was it sheer chance, or were there hidden laws at work? In other words, was it entirely impossible to foresee which of the young would survive in a given instance, or were there certain factors making it theoretically possible to predict who the "winners" would be?

Thinking about it, one becomes quickly aware that the assump-tion that pure accident decides the outcome of the struggle for survival contains an extremely improbable precondition: that all competing individuals have a fundamentally equal chance. Only with this prerequisite would it make any sense to speak of random results.

But fundamental equality can only be thought of as an ethical demand, not as a natural condition. No earthly power can change the fact that a handicapped child starts off with poorer chances than a healthy child of the same age. It is just *because* of this unequal opportunity, which also exists among healthy children, that we have the moral responsibility to protect, as far as we can, those affected by it.

And so in a bird's nest there is naturally no equality. Even at the moment of hatching all the nestlings do not have the same chance of becoming the two survivors who will replace the parents three years later. To be sure, sheer accident does play a role in the fate of the individual, and for this reason alone no scientist could confidently predict the outcome of an experimental study in this area, however carefully controlled.

But here too the chances aren't equal. From the first moment physical forces play a part: the most energetic beggar will likely have the fullest belly. Then too behavioral peculiarities are important. The fledgling that ducks down faster when a strange shadow appears by the nest increases its chance of survival. Not all young birds can register the presence of "something strange" with the same speed and accuracy.

In the course of later life these individual behavioral differences will become ever more crucial, at least among the higher animals. (In lower organisms, by contrast, physical differences are more fateful, while on the molecular level, naturally enough, only features of the material structure figure in the outcome of the contest.) To take one of literally countless examples in higher animals, consider the consequences of the genetically determined balance between the innate behavioral programs for curiosity and fear: the animal should not be too cautious, otherwise it will fail to have experiences of possible life-and-death importance or will have them too late. But it should't be too susceptible to curiosity either; a young bird overcome by curiosity the first time it meets a cat has already gambled away its chances.

Darwin came to the realization that individual differences of this sort play a critical role in deciding whether an organism will remain alive long enough to have offspring to take its place. This thesis, extremely simple but logically incontestable, laid the foundation of the theory that turned biology upside down.

Darwin's theory shows that the factors resulting from the inequality of individual chances determine the fate not only of each individual but, inevitably, that of the entire species. The population or reproductive community made up of all individuals of the same species begins to change over the course of generations. The change is slow and imperceptible to the human eye, but it comes on with irresistible force. No biological population is capable of resisting (without alteration) the consequences of the selection process that works on all its members through their varying degrees of fitness—this is Darwin's great discovery.

With the advantage of hindsight all this looks so simple that we wonder why nobody stumbled onto it before. But it's one of the peculiarities of the human psyche that it always takes a genius to discover the simple truths in the tangle of theories, opinions, and prejudices that baffle our minds—even though once these truths are brought to the light of day they immediately strike us as self-evident. The same is true here.

Let's summarize the situation one more time: of the 30 (or 5 or 100) offspring of a given pair only 2 will become parents in their turn, with the same number of offspring (at least if the population remains constant). What decides, apart from accidents, *which* of the 30 (or 5 or 100) offspring reach parenthood are differences with respect to specific qualities, that is, individual traits or characteristics.

Once again we can ignore the precise nature of these characteristics that play such an important and truly fateful role. Let's consider only the principle, the logic as it were, of the situation here and call the feature it has produced simply "X," without bothering our heads about its precise makeup. We shall merely assume that X is genetic in nature, that is, a hereditary quality.

So we have 2 offspring that have lived long enough to hatch 30 eggs. They have survived because of a hereditary character X, which distinguished them from all their siblings. But this assures us that X in the next generation (among the nestlings that will hatch from the eggs of the surviving pair) will necessarily be represented in greater numbers than in the parental generation.

The successful offspring, then, constitute a fundamentally new

hereditary "type," however imperceptibly small the difference
between them and others of their species. We can already see
which way things are headed: the competition for survival (Darwin, as everybody knows, called it the "struggle for existence")
will begin afresh in the new generation. Once more a selection
of 2 out of 30 will take place on the basis of certain hereditary
characteristics. In the next generation the genetic difference will
become a bit more marked; the population has begun, quite
slowly to be sure, to change its hereditary makeup.

And so we have, in broad outline, the mechanism that Darwin
found to be the cause, the driving force behind specific change:
a surplus of offspring in each generation with a correspondingly
high rate of loss; the selection of those few individuals as parents
of the following generation on the basis of certain genetic
peculiarities; the eugenic effect of this spontaneous natural process, which leads to the emergence of new species, by means of
the resulting gradual shift, over generations, of the hereditary
characteristics typical of a given population.[40]

Darwin thus rediscovered in the wild exactly the same principle
that human breeders had instinctively followed since the beginning of civilization. The domestic animals we know today are all
derived, without exception, from wild species, after our primal
forebears began to seek out and breed individuals whose physical
and temperamental qualities made them seem especially appropriate for human use.

If you capture wolves, raise them, and over a period of several
centuries choose the most biddable young for breeding, you
ultimately get the dog. Later in the history of civilization this
pattern of selection was determined by a whole variety of consid-
erations beyond practical need, such as aesthetic motives or the
typically human desire for novelty.

There can be no question that this is how the species dog came
into existence, along with all its many "artificial" breeds, from
the mastiff to the terrier, from the dachshund to the collie. And
there is likewise no doubt that fantail goldfish and the brilliantly
colored ornamental birds, like the incredible multiplicity of our

garden flowers, owe their existence to a similar eugenic selection of certain individuals for reproduction.

Nobody doubts the reality of "artificial" selective breeding or its results. But many people still can't grasp that natural selection could be the cause of the innumerable species that now populate the earth. Why not? Where's the difference?

Well, it does exist, but it strikes many people as *so* basic that they view the link between artificial and natural selection as dubious. The point is that in the first kind human breeders operate with a conscious goal or plan, while in the second there is no observable personal source to which any such plan could be attributed. Failing this, a number of people would argue, natural selection must be governed entirely by chance. But that would invalidate Darwin's explanation, because chance could never bring about the complex order we undeniably find in animate nature.

We have to take this objection seriously and hence address the question of who has been doing the selecting, and for what reasons, in the parts of nature free from human interference.

8. Accidental Order?

How LONG would it take a bunch of apes aimlessly pounding on typewriters to produce a single line from a sonnet by Shakespeare? Or, how long would you have to wait for a gust of wind to blow a pile of separate letters into a meaningful sentence? This argument is forever being brought up, in countless shapes and guises. And anyone who tries it on a lay audience can confidently expect applause. The argument in itself is logically consistent, and so it makes quite an impression—people like to hear this sort of thing. Among scientists, however, it's a dud. Despite its conclusiveness it has a central flaw—it has nothing whatsoever to do with the state of affairs it's supposed to refute.

True enough, the horde of apes would never manage to turn out that one line of Shakespeare. And the wind can't write, any more than—to take one more popular version—a heap of metal atoms could be randomly shaken until they became a Volkswagen.

But what does this mean? Certainly not what the anti-Darwinists who raise such objections think it means. All these images and metaphors simply give drastic expression to the truism that order can't arise by chance. That proposition is perfectly correct, but no one ever denied it, neither Darwin nor any other modern scientist.

We have already mentioned that modern science defines the emergence of order as the deviation from a random distribution. How then did the notion ever get abroad that Darwinism involved a kind of randomness which flies in the face of all probabil-

ity? This misunderstanding deserves an entire chapter, not only
for the sake of setting things straight, but also because the arguments and examples presented here will give us some idea of the
profound insight into nature that the current theory of evolution
affords.[41]

"Chance" is an ambiguous word that stirs up all sorts of confusion. It means, among other things, the lack of any order. It
denotes the opposite of meaning or a regular, recognizable pattern, and hence disarray, absurdity, unpredictability. This, and
this alone, is what the critics have in mind who feel obliged to
reject Darwinism because it unquestionably contains elements of
chance.

But there's much more to chance than these negative aspects.
Chance, for example, is related to freedom. We apply it to events
or processes when we have reason to believe that they have not
been regulated (determined) by some law.

If there were no chance in the universe, this world would be
nothing more than a gigantic machine running by fixed rules. We
could reconstruct the past in an unbroken line all the way back
to the beginning, and we could predict every detail of the future
till the end of time. Freedom of the will, historical responsibility,
and law would be illusory, truly superfluous concepts, since the
world would be bound by an unbroken causal chain that would
leave no room for freedom, without which the claims of morality
could have neither meaning nor necessity.

There are, we know, philosophical extremists who maintain
that this is in fact the case. And in an earlier epoch scientists, or
rather natural philosophers, sketched out the possibility of,
among other things, an ironclad predetermined world. Recall the
image of Laplace's *daimon,* who could have described every instant of the past and predicted every instant of the future
throughout the universe, provided only that he knew the position
of all its atoms at any one point in time.

Since then even the physicists have given up believing in such
a possibility. The first step away from it was taken by Werner
Heisenberg with his famous "indeterminacy principle." This re-

fers to the discovery that it is fundamentally impossible to specify simultaneously and precisely the location and momentum of an elementary particle. This did not come about, Heisenberg showed, as a result of some methodological problem or other in observing the subatomic realm but arose as a matter of principle; that is, we cannot even clearly define what sort of statement we would get in a simultaneous determination of the position and velocity of an electron or another particle. In the final analysis this derives from the peculiar (and literally unimaginable) hybrid nature of these elementary building blocks of matter, which can only be described as part wave and part particle.

Since in this case the initial conditions affecting the particle cannot be determined (and in a certain sense aren't even given), it is likewise fundamentally impossible to calculate in advance the future behavior of such a particle. Laplace's *daimon* has had the basis of his prophetic powers cut out from under him.

For a while physicists believed that this discovery left the causal determination of the macrocosm untouched. In all macrocosmic events such a vast number of elementary particles come into play that in this sphere—which includes our world—predictability can be reintroduced, on a secondary level as it were, by means of statistical averages. The oft cited comparison to a life insurance company illustrates this idea: the agent who has to figure out what premiums to charge has no way of knowing exactly when her client will die (and thus what his aggregate contribution to the policy will be). Nevertheless the agent can calculate the premium owed down to the penny, if her company has enough customers to enable her to use actuarial tables.

But this theory of "secondary" determinability did not hold up for long. Manfred Eigen demonstrated that random events on the molecular level can generate deviations in the macrocosm. Some years ago the Austrian physicist Roman Sexl devised an especially graphic picture of this situation when he calculated that if you sent a series of billiard balls caroming one into the other, even under ideal conditions, because of Heisenberg's uncertainty principle the seventh ball would not be guaranteed to hit the eighth. The uncertainty resulting from the indeterminate position of the molecules on the balls' surface, once raised to the eighth power, would equal one whole diameter of a ball.[42]

In 1977 the Belgian physicist Ilya Prigogine got the Nobel prize for proving that the random processes whose existence Darwin had guessed in a brilliant act of intuition more a hundred years before are also at work in physics, even in the macrophysical dimension. There too evolution has now become a critically important concept. Today we can see the deterministic model of the universe for what it was: a result of "excessive idealization" in classical mechanics, the "foundational myth" of classical science.[43]

The physicists have deprived Laplace's *daimon* of his cosmic authority in the atomic world. More than that, they don't even consider him an unbeatable billiards player. This does not, of course, *ipso facto* demonstrate the existence of free will, but at the very least human responsibility is once again an open question.

What concerns us here in this brief glimpse of the history of physics is the role played by chance in the course of things. At first it was banned from the universe, by order of the physicists, and the cosmos petrified into a gigantic clocklike mechanism, spinning absurdly along its predetermined path. To the degree that chance was once again admitted into the system and the course of the world was viewed as subject to incalculable forces, not chained to the laws of causality, the future of the cosmos became, once more, an open-ended one. It became a future *not* determined in advance, a future for which we have real not illusory decisions to make and in which we must prove ourselves (in which we can also come to grief.)

Once the openness that chance creates is banished from the world, there can be no doubt that decisions, responsibility, and moral law also vanish as pure subjective illusions. Where law is the absolute ruler, freedom can no longer exist. All those who narrow-mindedly equate chance with absurdity ought to reflect that the world would lose its meaning if chance played no part in it.

But the comparisons to the horde of apes or the sentence written by the wind have nothing whatsoever to do with Darwinism. Darwin never claimed that order could emerge from pure chance. What did he actually say? We must take a closer look at the role Darwin assigned to chance in evolutionary theory.

With peculiar obstinacy anti-Darwinist critics tend to overlook or conceal the fact that chance alone does not reign supreme in the theory of evolution. If chance were the sole driving force behind biological development, not one single functioning organism would have ever come into existence. Right from the start everything would have been doomed to end in total chaos. No biologist needs to be told that.

The crucial point of Darwin's explanation is the collaboration of random elements with natural laws. Chance by itself creates meaningless chaos. Law by itself creates automatism equally devoid of meaning. But together they prove to be, as Konrad Lorenz once put it, "the two great engineers" of specific change. To quote the subtitle of Eigen and Winkler's book, "The laws of nature guide the course of chance."

In evolutionary theory chance is represented by the principle of mutation. Law comes into play through the principle of selection, which follows hard and fast guidelines. Both principles call for some clarification.

We have already explained what a mutation is: a "mistake" in the copying (which must occur with every cell division) of the blueprints located in the nucleus. When a hereditary molecule, which constitutes a precisely defined "model," has to be copied millions of times over thousands or millions of generations, then "transmission errors" are quite unavoidable, regardless of the perfection of the genetic copying mechanism.

The whole process *is* close to perfect, as shown by the extremely low rate of error. It has to be very low, because otherwise the "memory of the species" (see the beginning of Chapter 4) couldn't fulfill its conservative mission of faithfully transmitting the blueprint for the organisms of a given species over the course of millions of years. On the other hand, there have to be some mistakes; hereditary transmission can't be absolutely flawless, because absolute conservatism here would mean an absolute standstill. If the genetic mechanism for reduplication functioned perfectly, then the earth would be crammed with identical copies of the first molecular system that succeeded in reproducing itself. Through all time, till the very end, there would be a boundless, ever increasing sameness.

It is clear that the rate of mutation (average number of errors per copying operation) is a critical factor in the history of life on earth. An increase in the rate would undoubtedly hasten the course of evolution, but at a certain point it would begin to jeopardize all further development because too much experimenting would be going on in each generation. A species that till then had been relatively stable but whose rate of mutation suddenly speeded up would bring forth within only a few generations such an abundance of crazy variants, monsters, and misbegotten creatures that it would soon die out from an excessive loss of genetic tradition. This seems to have been the fate of some of the species that disappeared from the face of the earth in the dim primeval past.

On the other hand, too low a rate of mutation would make the species extremely conservative. In exceptional cases this can be beneficial if it occurs in a species optimally suited to environmental conditions that are themselves conservative, that is, remain constant over geological epochs. Some species of cockroaches today look exactly like their ancestors hundreds of millions of years ago and there are other examples of such "living fossils."[44]

As a rule, however, extreme conservatism in nature proves to be as lethal as wild variety. An abnormally low rate of mutation presents the species with an insufficient number of alternatives, which it will need whenever changes in its environment call for a shift in genetic adaptation.

Hence, each one of the two principles in itself, both random change and the tendency to flawless copying, would in one way or another bring a species to disaster in no time at all. But together they strike a compromise, expressed in the rate of muta-

tion, and this is the first reason why living forms were so successful in taking over the earth.

The second reason is the principle of selection. Mutations by
themselves, however judicious their number, are not enough.
Because a mutation *per se* is not only random, it also has no
meaning. Its significance is only decided by the selective evaluation of the encounter with its environment. We shall have to
examine this connection in a little more detail.

Mutations are random in two ways. First of all, insofar as the
disturbance of the copying process that they represent takes
place on the atomic level. The blueprint being modified at some
point by a mutation exists in the form of a DNA or RNA molecule.
The insertion of a "false" base and with it the change in coding
at a given point of the molecule are arranged by processes that
occur at the level of elementary particles. For example, in the
case of natural radiation, which is the critical determinant of the
natural rate of mutation, the particles are helium nuclei, electrons, and photons. Since the behavior of such particles is undetermined (cannot be predicted or calculated in advance), mutations are fundamentally random events. True, a molecular
biologist can cite certain probabilities for a mutation occurring
at this or that point of the molecule. But there is no way of
foreseeing when this will happen or what the mutation will consist in (which link in the original molecular chain will, when
copied, be exchanged for which new link).

But within the framework of evolutionary history mutations are
random events in an entirely different and more basic sense—
they take place with no regard for the situation of the population
whose gene pool they alter. Whether the composition of this
gene pool (the sum of all hereditary structures of the individuals
making up the population) is ideally adapted to the prevailing
environmental conditions or not is of no consequence. Even in
cases of optimal adaptation, mutations will not decrease in frequency. Nor will they increase when the species has an urgent
need of them, should drastic environmental changes make ac-
celerated genetic adaptation desirable.

We can readily see why that sort of response, however convenient it would surely be, is altogether impossible. Such a tie-in
between the rate or direction of mutation and the need for it is
out of the question because rate and need have entirely different
causes, which operate in widely separate regions of nature.

Biologically significant changes in the environment consist in,
for example, a long-term climatic alteration, the causes of which
may be astronomic (such as fluctuations in the sun's activity) or
geological (changes in the atmosphere's radiational permeability
owing to volcanic eruptions) or human (clearing of virgin forests,
increased concentration of carbon dioxide in the atmosphere).
They can lead to a change in the vegetation and hence of the food
supply or to the appearance of a new competitor for the available
resources, perhaps as a result of animal migrations, which may in
turn arise from climatic or geological causes (territorial displace-
ments owing to floods or the emergence of mountains).

The factors determining the rate or kind of mutation, however,
are altogether different. They all derive from the atomic, microcosmic realm. They are, in the first instance, physico-chemical
forces that control the stability of the hereditary molecule at the
various points of its structure: binding forces on the surface of
the different atoms that compose it, or the physico-chemical com-
patibility of the electron shells of neighboring links in the chain.
These are the sort of factors that decide under what conditions
the molecule could lose one of its elements at a certain point or
replace it with a new one.

There's no need to explain at length why these factors as a
whole are independent of the environmental factors to which an
organism must be adapted (within certain limits) if it is to survive.
The biological requirements of a given climatic situation, of a
given predator–prey relationship or of a given food supply exist
on a level that is in no way connected to the physico-chemical
forces that determine the mutation of an RNA strand. There is
simply no causal tie between them—such a tie is not even
theoretically conceivable.

Thus a mutation is random not merely in the sense that it can't be predicted but also insofar as it occurs with no reference to the biological needs of the organism whose hereditary material it alters. It is blind to the biological situation it helps to decide. Isn't this state of affairs the quintessence of aimlessness?

But here too nature has found a way to wrest a meaning (retroactively) from the whole process.[45] True, the genome (or sum of an organism's hereditary tendencies) can never learn from experience, can neither acquire information from the environment nor profit from mistakes in mutational adaptation. No information from the world of organisms ever penetrates into the world of elementary particles. But the resultant blindness of every individual mutation to the situation of the organism with whose blueprint it mindlessly tinkers has nevertheless an advantage: this unavoidable blindness leaves the species open to unforeseeable future possibilities.

We have already cited as an example of the genome's "unteachableness" the fact that the most varied species keep on producing albinos, a mutation that under normal circumstances would bring nothing but harm to a deer, a blackbird, or a mouse. In nature these variants are very quickly "selected out" by their environment. If they don't fall prey to their enemies more readily because of their conspicuousness, then they may be so busy saving their skin that they don't have enough time to raise young. And that would be the only way that the new gene for albinism could become a permanent feature of the gene pool of the species in question.

To that extent it makes no sense that the blind mechanism of mutation should keep on producing the same albino variant over millions of years. But under certain circumstances the wide-ranging fantasy of the mutational principle, which pays no heed to the realities before it, can suddenly prove to be a lifesaver for the species. This happens when environmental conditions change so unpredictably that a mechanism for creating mutations which was totally absorbed by the concrete demands of the present would be wiped out. In such situations (and in all probability they have continually arisen throughout geological history) a random shot can suddenly become a bull's-eye no marksman would have managed to hit because there seemed to be no reason to aim in the direction that later turned out to be the right one.

Polar bears, ptarmigans, and snowshoe hares have this fact to thank for their existence. They are all the descendants of albino variants that emerged—through blind chance—during epochs when their populations were driven into a permanently snow-covered environment for whatever reasons—stronger rivals, large-scale climatic change (ice age), or geological catastrophes. At that point the white variant, which had arisen and persisted without serving any particular purpose, abruptly showed itself to be very useful indeed.

This turn of events—let me stress the fact once more, important as it is for understanding evolution—was neither foreseen (the genome can't "aim"—mutations occur haphazardly) nor foreseeable (ice ages, geological catastrophes, or the invasion of superior competitors are not announced in advance). All we can say is that the lucky hit may, under these circumstances, have been made possible only by a totally blind "potshot" through the mutational process. Because a marksman, as we said before, might well have deliberately not fired in the "right" direction, since at the moment when he pulled the trigger there wouldn't have been anything to aim at.

Perhaps in the case of an ice age our marksman might have been equal to the situation, providing that such periods of cold weather really did move in gradually over the course of many thousands of years (some theories call for much shorter time spans). But in the vast majority of cases the process of evolutionary adaptation, which by our standards works very slowly, would never have had a chance of catching up with changes in the environment if it simply reacted to them instead of accidentally anticipating them.

Which immediately raises the next question: to what extent can accidental mutations in fact anticipate unforeseeable changes in the environment? Or, in other words, how great is the likelihood that such a completely blind operation could cope with unpredictable future changes? For many nonbiologists opposed to the theory of evolution this is a purely rhetorical question. They are convinced that here, if not some place before, Darwinism has finally run aground. The objection may sound plausible, but biology has a solid answer ready. Still, let's hold it in abeyance for a moment while we go back to the relationship between mutations and natural selection.

It should be clear by now that the value of a mutation can only be judged after the fact. It has likewise become apparent that the environment in which the organism must survive has the final say in this matter. We recall the "trifling individual differences" between various specimens that Darwin postulated as the starting point of his explanation. In his day nothing was known about the genetic code or mutations. Today we know that these individual differences really exist, and that they can be traced back to to mutations in the germ cells and to their subsequent intermingling thanks to bisexual reproduction.

Whether a mutation is to get a positive or negative evaluation depends, speaking in Darwinian terms, upon whether it increases or lessens the individual's chances to be one of the select few members of the next generation of parents. And whether or not this happens is something that we can never tell simply by looking at the mutation itself. The decision rests with the effects that the mutation has on the organism's performance in the environmental test. (An albino's white fur or feathers can't be rated as a positive or negative quality until the possessor of such fur or feathers is transplanted into an environment where the new characteristic has positive or negative results, depending on the circumstances.)

One more example: it seems, at first blush, utterly pointless that a living creature should undergo a specific mutation causing it to rearrange its metabolism so as to squander a significant part of its food intake to raise its body temperature above that of the environment. For hundreds and hundreds of millions of years such a mutation would have been doomed to failure. During the vast time period in which the only life on earth was in the water any effort in this direction would have been nipped in the bud. In an oceanic environment heating one's own body makes no sense at all. Just a few meters beneath the surface of the sea temperatures remain constant year in and year out. Here it would be highly disadvantageous to expend a large portion of one's nourishment for a worthless function.

Hence, as soon as this mutation appeared (and presumably it did so again and again, owing to the blind mechanism that produced it), it was immediately eliminated by negative selection. This most likely occurred when the individuals who had undergone the mutation developed an increased need for food and thus fell so far behind their (still cold-blooded) fellow creatures that their chance of passing on the new hereditary variant to their offspring was drastically reduced. This is what biologists mean when they talk about selection by the environment.

Nevertheless, it would be false and short-sighted to consider such a mutation fundamentally negative. The "squandering of nourishment," which seemed at first so pointless, proved to be highly advantageous at a much later stage of the earth's history. The criterion changed when the environment likewise underwent an unforeseeable alteration: life gradually began to encroach upon dry land.

It appears (for reasons based on the physico-chemical structure of the hereditary molecule) that the mutation for warm-bloodedness occurred with relative infrequency. We can draw this conclusion from the fact that after living creatures emerged from the water a long period of time passed before the mutation began to spread in certain populations. The first animals ventured out onto terra firma around 400 million years ago, but not until 250 million years later did the rudiments of warm-bloodedness show up in a number of saurian species.

Today the earth is ruled by the heirs of a mutation that was originally senseless and that therefore remained a failure for almost half a billion years. One might be tempted to say that the blind unteachableness of the process that produced it was actually fortunate. A creative mechanism capable of storing up experience, after such a vast amount of time with exclusively negative results to show for it, might well have given up repeating this mutation before life began to adapt to conditions in the open air.

There—and only there—the same variant suddenly proved advantageous to its owner. The higher consumption of nourishment was more than compensated by what the organism gained in independence from the temperature variations typical of the new milieu. The poikilotherms (cold-blooded creatures) whose dominance of the earth had gone unchallenged up till then—think of the 150 or more million years that the saurians reigned supreme—unexpectedly found themselves up against a competitor whose activity was not subject, like theirs, to the rhythm of external temperatures. We all know what the consequences were. The most favored heirs of this "thermal emancipation" are we ourselves, the human race.

Thus we can legitimately compare the relationship between mutation and selection to that between a doughlike material and the form that molds it. Mutations are what give a species the capability for genetic adaptation ("kneadability"). Their very aimlessness and randomness make it possible for adaptation of almost any conceivable kind to take place. (Only up to a point, of course, because among other things already existent patterns put limits on further development: a horse will never turn into the winged Pegasus, and we will most certainly never be able to breathe through gills.)

But there is more to the comparison. Too many mutations would make the structures of a given species too "soft." They would break down and dissolve. Still, in addition to this a formative power is required. Random mutations by themselves are incapable of producing an organism; they are rather the indispensable prerequisite for that production, which can only be realized by the intervention of the environment, selecting among mutations in accordance with definite criteria. Nobody has ever assigned any other role to chance in evolution—one wonders if any of the people invoking the ape-horde argument will ever realize this.

Under these circumstances order fits into the evolutionary process because the environment always contained ordered structures. This holds as a matter of principle, from the very beginning. It holds for the origin of galaxies, planetary systems, and suns from the chaos of the cloud of radiation released by the Big Bang. The only reason these cosmic patterns could come into existence was that constant natural laws and fixed structures of the atoms involved were at work establishing order. One often hears the objection that the laws of thermodynamics (the principle of entropy) would have only destroyed order, never have allowed it to arise.[46] But even in this context that argument just doesn't stand up.

Order, therefore, as represented by complex environmental structures, works as the formative power, shaping the plastic substance of the available mutations. The incredible abundance and complexity of living species present on earth thus reflects the unimaginably vast number of different ways of discovering and exploiting ever new environments on the earth's surface: new constellations of matter keep arising, which vary ever so slightly from previous ones, and come up with combinations of natural conditions that no other competitor ever made use of.

This principle explains why air, which seems such a monotonous, characterless element, could provide the setting not only for bees, butterflies, and darning needles, but also for birds and bats and countless other types of "fliers." And if we reflect that the term environment is obviously not limited to inanimate external factors, but includes all the other living creatures (and their behavioral peculiarities) in an organism's environment, then we realize that evolution is a self-reinforcing process.

By the same token, in bringing forth a continually new stream of forms and structures, it exponentially increases the complexity of the environment and therewith the number of possibilities for future adaptation. This would help to explain the acceleration of the evolutionary process that has been observed over the course of time.

The interplay between mutational plasticity and formative environmental influences is extremely close. This can be seen with brilliant clarity in certain exceptional cases, which I should like to mention briefly. There are special adaptive forms that owe their peculiar character to the *disappearance* from their environment of one of its original features. The best known example of this is provided by certain cave dwelling creatures (cave salamanders, fish, spiders, and insects), which have not only become blind but whose eyes in many cases have undergone a more or less complete regression. Since these animals are without exception related to species that normally live in daylight and have excellent vision, we can assume that some millennia ago these populations were trapped by geological events (landslides, a rise in the water level, etc.) in the sunless cave environment where we now find them.

How remarkable that this circumstance has led to their loss of vision and even to retrogression of the eyes. The logical conclusion is that not only is the environment needed for new hereditary characteristics to emerge, but its permanent collaboration is required to maintain them. The species' genetic plasticity submits to the influence of its environment even when that milieu drops one of its conditions for survival.

How are we to imagine the procedures followed by natural selection in such a case? If the concept of selection is valid, we have to assume, in the face of the cave dwellers' genetic destiny, that the mutation of eyelessness or blindness evidently constitutes an advantage in total darkness. That doesn't sound very plausible at first—where could the advantage lie?

From the very beginning biologists suspected that the answer had to derive from the principle of economy in nature. When the eye no longer serves any purpose, the expenditure of energy needed for the upkeep of that organ becomes a pure nuisance. Any creature that manages to get rid of this now useless expense by an appropriate mutation will therefore live more economically and efficiently than its rivals. On first hearing this explanation must strike many people as somewhat farfetched. Yet it not only fits in with the theory of natural selection, it is also quite obviously correct—as was proved by a highly interesting experiment conducted with mushrooms.[47]

Unlike true plants mushrooms can't manufacture any of their necessary molecular building blocks with the help of solar energy, through photosynthesis, out of simple compounds. They lack the chlorophyll required for this operation. Instead they have to take in many of the substances they need "ready-made," mostly in the form of dead plants or animals. This is what makes mushrooms typical saprophytes (from the Greek *sapros*, "rotten").

This peculiarity served as the foundation of the experiment. Scientists worked with two mutants of the same species of mushroom, genetically identical with one exception: the first sort was capable of producing by itself a certain amino acid; the other had to get it ready-made from its environment in order to survive and reproduce. Hence the innate capacities of the first sort (let us call it type A) were greater than those of the second sort (type B). With respect to a substance that was vital to both A and B, type A was self-sufficient, type B was not.

One would thus assume that in evolutionary competition type A would have to beat out its less gifted rival. And under ordinary circumstances that would most likely happen. Type A simply has a better chance of landing in a soil where it will thrive than type B, which can only survive in an environment that in addition meets the condition of supplying the amino acid which B itself, unlike A, can't produce.

But the experimenters did not allow both types to compete under ordinary circumstances. They put them instead in an artificial substrate that contained everything they needed, including the amino acid that A had no need for, owing to its greater synthesizing capacity. The experiment only lasted a few days, but the results were decisive. Within this short time B completely routed its apparently superior rival. The surface of the substrate was positively swarming with B, while A had vanished.

The experiment was repeated, under the same conditions, and each time it produced the same result. In this substrate B was enormously superior to A. Why? The only possible answer was that A got the worst of it because its genetic program compelled it to invest energy in a capability that was superfluous in this special environment. That was enough. The apparently minimal difference between the two competitors, that is, that one had to waste a tiny portion of its energy, was all that natural selection needed to gain its decisive leverage. (It goes without saying that this hypothesis, though the nearest at hand, does not exclude the possibility that still other—and perhaps even more powerful forces—played a part in the final result.)

There is a clear parallel here to the fish that a geological accident trapped in the eternal darkness of a cave. Once again we see the thorough intertwining, the perfect correspondence of mutational adaptability on the one hand and the formative power of the environment on the other.

This perfection of the correspondence between the supply of mutations as shaped matter and environmental selection as shaping form provides the background for the mimicry characteristic of all biological adaptations: the fish's fin "copies" the water in the same sense as the horse's hoof copies the ground out on the steppes, and the ape's grasping hand copies the branches of a tree.[48] The fish couldn't swim, the horse couldn't gallop, and the ape couldn't climb unless this were so.

Nor would we be able to see anything, if millions of years ago our eye had not "discovered" and copied (in its structural principles and mode of functioning) the optical qualities of the earth's atmosphere and of refractive mediums, so recently and laboriously analyzed by science, along with the spectral composition of sunlight—if our eye, as Goethe said (anticipating Lorenz), were not "sunlike."

All this is extraordinarily astonishing, but it's also quite understandable. We should also certainly marvel that, without being entitled to it, we are undeniably in a position to discover and, at least to some extent, comprehend these connections. We have even managed in the foregoing survey to dispense with any appeal to a plan or goal of the evolutionary process. So far the

dialectical interplay of mutation and natural selection has satisfactorily answered all our questions without recourse to some
power or authority drawing nature forward from out of the future
or prescient of future goals. But this in no way settles the problem of the relation between evolution and the principle that we
call "mind," a question we mustn't ignore. But before we can
discuss it intelligently we have to do a little more spade work.

To close out this chapter, how plausible in fact is the assumption of modern evolutionary theory that random mutations anticipate wholly unforeseeable conditions of adaptation, which lie in
an open-ended future? The answer is, we don't find this assumption completely validated but far more so than many consider
possible, and far more than seems believable to our limited
imaginations.

As we would expect, haphazard, scattershot mutations cannot
anticipate every future possibility. The proof of this is that whole
species become extinct. Otherwise the world would still teem
with ammonites and saber-tooth tigers, with woolly rhinoceroses,
mammoths, and countless other species. These creatures disappeared because one day they came up against new environmental
circumstances (of whatever origin) to which their species had no
ready response. They encountered situations where their
anatomical structure and behavioral repertoire proved to be in-
sufficiently adaptive, since the species failed to produce, through
random mutations, the hereditary variants needed at this critical
juncture.

Intuitively we view it as highly improbable that future needs
could be met by pure chance. And so we're not surprised to learn
that extinction is an extraordinarily frequent event in evolutionary history. It is a rule among students of evolution that no
species can go on existing indefinitely. Without any doubt the
number of extinct species is many hundreds of times greater than
that of currently living species.

We have no difficulty believing any of this, but we also have to
look at the other side of the coin. A completely random anticipation of future environmental demands can't be ruled out. This

does occur far more often than we would think likely—as shown by the indisputable fact that most species last a good long time. Their geological life expectancy is in the range of millions of years. One could assume that environmental conditions remained constant over such a long period, but only in exceptional cases would this explanation work.

There is another point that must be considered here. The saurians, to take one example, did indeed vanish from the earth, but their line didn't actually die out, insofar as they still have living descendants today (birds, among others). So we have one class of creatures that disappeared not from becoming extinct but from undergoing a momentous transformation, which allowed at least some of its species to go on living in a new form. Thus we really should count such cases as one more accidental bull's-eye.

The success rate of the mutational process, as a matter of fact, greatly exceeds our intuitive expectations. This was conclusively proved by an experiment as ingenious as it was simple, devised by the American Nobel prize winner Joshua Lederberg. His starting point was the clinically tested fact that pathogenic bacteria can develop a specific resistance to antibiotics. This can be easily observed in the bacteriological lab: if you expose a bacterial culture to a deadly (for the culture) poison, such as streptomycin, all the pathogens usually die. Sometimes, however, it happens that at a point in the culture medium, which has been swept clear by the streptomycin, a tiny colony of bacteria begins to grow anew until finally within a few days it blankets the bottom of the Petri dish. What we have are the descendants of the pathogenic strain that have suddenly proved resistant, immune as it were, to streptomycin.

The whole thing is a textbook case of evolution in the laboratory. Among the hundreds of millions of bacteria growing in the culture when the antibiotic was added to it there must have been a single bacterium somewhere with a random mutation protecting it against the poison. Naturally this mutation, in an environment altered by that very poison, created an overwhelming survival advantage. While all its fellows died, this one bacterium, blessed by chance, became the parent cell of all the succeeding generations that came into existence (through division) and repopulated the culture medium. None of them had any trouble whatsoever with the poison, because they all inherited the mutation from their one common ancestral cell.

Hence in this case a single individual caused the genetic alteration of its stock. The only question is whether the mutation that made this possible and increased its chances over against all the other bacteria was really a random event, as the theory insists. Is it really conceivable that the quite specific genetic changes, which give a bacterium its resistance to a certain kind of antibiotic could have come about by accident? Don't we have to consider the possibility that the cells may have "learned" their resistance through contact with the streptomycin (in the same way that we acquire a specific immunity as a result of, and in reaction to, contact with a certain kind of virus)?

Well, whether it's conceivable or not, Lederberg proved that in his experiment it had to be a random mutation. He doubled his original colony of bacteria by producing an identical copy of it on a second Petri dish. Then he added the antibiotic to the first dish and waited to see if a colony with resistant mutants would emerge at any point.

If that occurred, then he turned to the second dish, which was to give him the proof he was looking for. And, as a matter of fact, when he removed bacteria from the point in the dish that mirrored the spot where the resistant colony had developed, all these cells likewise turned out to be resistant. In contrast, bacteria coming from any other point on either of the two shells always fell victim to the streptomycin. None of them ever learned to cope with the poison.

Thus Lederberg had proof that at this one point, right from the start, there must have been a bacterium which even before its first contact with the antibiotic had become resistant to it. The experiment can be repeated with the most varied sorts of bacteria and antibiotics, and the result is always the same. Here we have evidence that a mutation can anticipate conditions that the cell, if it wishes to survive, will have to meet when it comes into contact with the (as yet unknown) poison.

The experiment also shows that this anticipation doesn't always succeed, not by a long shot. But neither is it as impossible as it always seems. In reality it's by no means a rare occurrence, as demonstrated by the fact that for many years now the spread of resistant strains of bacteria has been a serious medical problem.

It's enough to make one shake one's head in disbelief. This means, among other things, that thousands of years ago among the most varied species of bacteria mutations must have been continually popping up, which would have imparted a resistance to penicillin, streptomycin, or any other of the many modern antibiotics—if they had existed then. But at that time these mutations were doubtless quickly destroyed by natural selection, since they made (as yet) no sense.

And we can be just as sure that among the astronomical number of mutations produced every instant by present-day bacterial populations, there must be some that are resistant to drugs that do not yet exist. All this is hard, if not impossible, for some people to imagine. I too find it difficult. But we must not forget that whether or not we can imagine a thing is no argument where nature is concerned. It's also impossible for us to conceive that the universe is boundless and yet not infinitely large. Yet we have this paradoxical truth physically present, right before our eyes. We need only lift up our heads and look at the stars.

And yet, how arrogant of us to presume (even if silently) that all of nature's mysteries must fit within the narrow horizon of our imagination.

9. Notes on a Grisly Theme: "The Struggle for Existence"

DARWIN, as we recall, based his explanation of natural selection among individuals with unequal chances of survival upon the following principle: the phenomenon of species change results from the fact that in each generation those individuals are chosen or selected out whose physical equipment and behavioral potential best correspond both to the risks and possibilities of their environment. Thus the best adapted individual survives. Darwin called this "the survival of the fittest."* To describe the competition itself, which provides the framework for natural selection, Darwin coined the famous catch phrase, "the struggle for existence." Both terms were widely misunderstood, with devastating consequences.

When we register an objective state of affairs in language, there is no way to prevent irrelevant connotations from creeping in. All too easily these suggest certain conclusions that have nothing to do with the facts or even distort them. This situation is so unalterable that we generally pay no attention to it.

One of things that causes it is the anthropocentric structure of our language, which was not designed for empirical-objective description of the world. It developed historically, serving the expressive talents and communication needs of living subjects. In keeping with this origin the structure of language takes for

---

* The phrase was actually coined somewhat later by Herbert Spencer.—TRANS.

granted a perspective identical to that of the experiencing and linguistically communicating subject.

And so every evening as we observe the decreasing distance between the disk of the sun and the western horizon, we say that "the sun is setting," because the identification of language with the perspective of the observer precludes the notion that the subject too might be moving. Just how much this peculiarity of linguistic structure can bias our judgment is evidenced by the fact that it took an intellectual revolution to discard the popular notion that the sun moves around us and recognize the objective situation. And just how conservative such structures of language are can be seen in our unembarrassed use of the old expression today—400 years after Copernicus—even though we know better.

There are many other kinds of similar linguistic "prejudices." Our language is, among other things, syntactically laid out so that the subject of a sentence is always characterized as doing or experiencing something (unless the sentence merely states the subject exists). We say, for example, that a tree "rustles," although in reality the wind moves its leaves. (But, of course, even this expression is misleading, since it assigns an active role to "the" wind.)

These remarks are made in hopes of arousing a certain (highly advisable) mistrust toward the assumption that conventional verbal descriptions can be understood with absolute literalness when they refer to matters outside the realm of everyday experience. The gap between the meanings that can be grasped by ordinary language and the nature of the facts to be described grows all the wider as those facts recede further from the field of ordinary experience. This is why physicists, as many people know, were long ago compelled to work out an artificial mathematical language in order to do justice to the microcosmic reality that lies behind appearances.

Both these expressions, "the survival of the fittest" and "the struggle for existence," were seized upon by a society that believed in science. They were taken literally and dangerously misunderstood. For many people these handy phrases seemed to define univocally the "natural law" governing success in all areas of life. This law presented itself to them as a seductive formula for social and political success.

But anyone who takes "survival of the fittest" in this literal sense, misunderstanding it as as a revelation of nature that applies to all life including human social life, will claim before long that "might makes right." From there it's not very far to the concept of individuals who don't deserve, who have no right, to live. Or to the idea that certain races, cultures, or nations are fundamentally superior to others. In the end, all these interpretations lead to a standard of "values" (supposedly legitimated by nature), measured against which ethical norms and limits can easily be mocked as expressions of sentimental fantasy if not of cowardice.

So much by way of a reminder of the terrible part played by Social Darwinism, especially in German politics. There can be no doubt that the unsurpassable horrors and bestial crimes of the Nazi era had one of their intellectual roots in that pseudo-scientific philosophy of social life.

Isn't that enough to unmask Darwinism as a false and inhuman doctrine? Many people are so deeply convinced of this that it's hard to make them see how they have been taken in by prejudice. Hence, before I discuss the arguments for evolution in detail, I should like to try to show, by means of a historical parallel, why there are no admissible *moral* grounds for denying Darwin's theory.

The parallel situation is the horrible persecution of witches during the Middle Ages. While it lasted, hundreds of thousands of women and girls were brutally tortured and murdered in Western Europe, because in one way or another (usually through anonymous informers) they had come under suspicion of making a pact with the devil. Thus the motive for the persecution was religious.

To be sure, the conduct of the trial and execution of witches was in the hands of the secular power. But the *Hammer of Witches,* the manual that laid down the procedural rules (and that recommended, among other things, the use of "endless torture" until a confession was obtained), was written by two Dominican monks and approved by a papal bull in the year 1484. In that document Pope Innocent VIII proclaimed with solemn emphasis that the abomination of witchcraft was a real and present danger.

A similar pattern holds for the persecution of heretics, which in the Netherlands alone during the reign of Charles V is said to have claimed more than fifty thousand victims. Here too—and in all the other countries where the Inquisition raged for centuries —trials, torture, and executions devolved upon the secular authorities, but the ferreting out of suspected persons was the church's business. As early as 1215 a church council declared that this was one of the most important tasks of bishops—and this regulation was not cancelled until the nineteenth century.

The horrors for which our own age is responsible disqualify us from exploding in indignation over these historical facts. But, quite apart from that, only a malicious person would think of connecting these dreadful aberrations with the essence of Christianity. That wouldn't be a permissible conclusion even if it were certain that everyone taking part in a trial by the Inquisition had always acted from purely religious motives. It must have occasionally happened that judges and even, perhaps, executioners were honestly convinced that they were meeting a churchly obligation. More than that, some surely acted in the belief that they were doing the best thing for the offender by using what was supposedly the only means to save his or her soul from eternal damnation.

But even then, if they all felt themselves justified or obliged with reference to their Christian convictions to do what they did, we would not regard their behavior as a legitimate expression of Christian piety. Even then we would continue to condemn it as the result of an atrocious misinterpretation, a perversion of the real meaning of Christianity. Our judgment would not in the least be altered by the fact that for a long time this perverted interpretation was considered the right one by church dignitaries, including several popes, who embraced it with enthusiasm.

This is just how I see the relationship between the advocates of Social Darwinism and what evolutionary theory, going back to Darwin, really means. Strictly speaking, the misunderstanding begins with the very term "Darwinism." In general, we use the suffix "ism" to designate a conviction or value judgment, but Darwin's theory is a description of objective facts.

This distinction implies no disparagement. It's simply a matter of clearly separating different categories. After all, we don't talk about "Copernicanism" apropos of modern astronomy or of "Einsteinism" in the case of the theory of relativity. But, however that may be, we can't lay the blame for the perversion of Social Darwinism at the door of evolution, even if that doctrine had some renowned scientists among its defenders. People may mistakenly derive maxims for human relations from a scientific theory, but that's not the theory's fault.

But above all we must not take the term "struggle for existence" as literally as the Social Darwinists do. The phrase refers to the relatively complicated mechanism of natural selection, which has by no means revealed all its secrets to scientific research. As remote as this process is from our everyday experience, the literal meaning of the everyday expression describing it is just as remote from its real significance.

The first evidence for this claim is provided by Konrad Lorenz's observations on the Tasmanian wolf (or thylacine).[49] He shows that "competition from one's fellows is more deadly than attacks by one's deadliest enemy."

The handwriting on the wall appeared for the thylacine when the first immigrants introduced the dog to Australia. Many of these animals ran wild and became the ancestors of the modern dingo. The dingo lived in the wilderness as a predatory carnivore, feeding on smaller marsupials. To this day not one of these smaller animals has been wiped out by the dingo—but the thylacine has.

Yet this didn't happen in the bloody fashion imagined by those who take the "struggle for existence" literally. In the conflict between dingo and thylacine not a drop of blood was shed, but the thylacine was nonetheless eliminated. There was no bloodshed because the thylacine was so vastly superior as a fighter to its new rival that the dingo must have taken good care not to tangle with it. But the thylacine lost out anyway, because the dingo's skill as a hunter so far exceeded its own that its chances of survival quickly dwindled.

We don't even have to assume that so much as one thylacine starved to death. The connections between the two animals must have been more complex than that. It is more likely that, among other things, the thylacine found the job of procuring food under the new circumstances so tiring and time consuming that its sexual and family life fared badly, with the result that with each generation the number of its descendants shrank. That was all it took.

We can see with particular clarity how sophisticated the "struggle for existence" really is in the cases where individual life expectancy and success in natural selection are negatively correlated. This must strike the crude understanding of the average social Darwinist as quite a paradox, but it happens. Lorenz cites instances from the area of sexual selection.

In many species actual mating is preceded by a longish courtship during which the male tries to draw the attention of the female. Very often males have special visible features that serve this purpose. Scientists speak of regular display organs in cases where extravagant characteristics have developed which are of no value except in courtship and which, apart from this one function, actually put the individuals possessing them at a disadvantage.

This holds true for the deer's antlers, as well as for the exuberant plumage of the bird of paradise and the tail of the peacock. A stag would undoubtedly be better off without antlers, just as a cock pheasant would be without its lavish feathers. Both creatures would surely have an easier time surviving without these ornamental features. But, as Lorenz rightfully points out, they would also leave behind fewer offspring or none at all, since they would fail in the competition for a mate.

I can't repeat often enough that everything depends on the number of offspring. (It's the only way that a hereditary characteristic can survive and perhaps spread throughout the population.) In the final analysis reproductive success determines everything and not the immediate confrontation with a rival, at least not in the sense of mortal combat that annihilates one of the two competitors. Instead immediate confrontations, where they occur, are only a factor in determining an animal's chances of reproduction.

The battle between rivals for a certain female is only a ritualized conflict, a kind of (usually harmless) duel. Nature has even produced special modes of behavior (innate "submissive gestures") simply to enable such conflicts to be decided without risk of death. Anyone who still takes the "struggle for existence" to be a war of extermination of all against all should keep this in mind.[50]

If that were really how natural selection worked, then the earth today would be a chamber of horrors filled with bristling, murderous monsters and armor-plated, practically immobile dinosaurs. But it isn't. And the innovations that determined whether a species would succeed or fail did not consist in the development of ever sharper teeth and claws or thicker protective coats but in improvements of a far different and much more subtle kind.

Evolution unquestionably did give rise to talons and fangs, but these are not for killing members of the same species but are for self-defense, and especially the capture and killing of prey. Life in the wild is no doubt a less idyllic affair than we like to paint it, but do we really have the right to accuse nature of bloody cruelty? Of all living creatures on this planet there is only one which engages in the supreme brutality of murdering members of its own species, which attacks its fellows with the deliberate intention of destroying them—and that creature is man.

To this day the combined efforts of cats and birds of prey have failed to kill off all the mice. In the same way the capacity for killing other creatures (the supposed natural law that "might makes right") has little to do with the qualities that make a creature a good candidate for survival. And when competition between members of the same species leads to changes in that species, it is not for the most part head-on conflicts that decide the outcome of the "struggle for existence."

As a rule the individuals among whom evolutionary confrontation (the "struggle," in Darwin's sense) occurs do not even see each other. Günther Osche cites a graphic example of this.[51] He describes a "showdown" that determines which of two weasels is the fitter. Both are hunting at the edge of a forest and far enough apart so that neither observes the other, when high in the sky above them appears the silhouette of a hawk. One of the weasels is so carried away by enthusiasm for the chase that it pays no attention to the danger. The other, however, for all its eagerness, regularly stops and sniffs the air—and, noticing the shadow in the sky, it hides. The bird swoops down on the careless first weasel and kills it. In this episode the literal confrontation took place between the hawk and the first weasel, but the "struggle for existence" was really between the two weasels, even though they never met.

The decisive factor in the course of evolution is not physical weaponry but the much more elegant and momentous process of innovation. A better strategy for ensuring safety might be developed, as in the weasel example, or a new enzyme permitting a shift to a new food source untouched by competitors. Or there might develop some other mutation, seemingly useless at first, that one day, when environmental change occurs, will turn into a promising innovation. We have already discussed instances of this kind in the example of albinos and warm-bloodedness.

What the layperson must realize is that the enormous number of different species produced by evolution, from the gnat to the elephant, from the scorpion to the eagle, has come about precisely because of a clear-cut strategy of *avoiding* conflict. Wherever it's possible to occupy a new environmental niche, undisturbed to some extent by competitors, the pressure of natural selection will begin to press for adaptation to just this niche, this novel cluster of circumstances.

Thus the creativity of evolution is not the result of "decisions" reached through a concrete, permanent struggle, but quite the contrary—it derives from the prevalent tendency to dodge the pressure of competition and its potential conflicts by turning to ever new modes of adaptation. The incredible number of animal species that fill the earth today lets us infer how frequently this attempt must have succeeded.

In conclusion let it be noted that there *are* some legitimate parallels between the processes that occur in evolution and those that we find in the human cultural world. This applies above all to certain technical developments. The analogies here are so explicit that problems in technical development could be solved with the help of strategies borrowed from evolution.[52] But even in these cases the processes that lead to the introduction of a new variant and the extinction of a prior form don't in the least resemble the sort of conflict that the phrase "struggle for existence" admittedly and unfortunately suggests.

In the struggle for existence on the high seas steamships have unquestionably won; they have driven sailing ships to extinction —but not by ramming and sinking them. A technical innovation (or mutation), namely the discovery of the steam engine and the harnessing of coal as an energy source, gave rise to a new type of ship. From then on this type was preferentially selected (ordered by shipping companies) because it was better adapted to the demands of worldwide maritime commerce.

But who knows whether this is the end of the story. A further change might occur in the environment, the supply of fossil fuels might dry up, and then the criteria of (economic) selection might be turned upside down. We might wind up in a renaissance of the wind-powered ship (no fuel bills), which has only survived today in certain adaptive species, chiefly pleasure craft. Thus there are no absolute goals, even in technology, of the sort that are permanently anchored in advance, toward which we can chart a methodical, unvarying course. In technology as in everything else the environment decides.

10. False Prophets

IT'S ALMOST impossible to keep track of all the objections that outsiders keep bringing up against the theory of evolution, but the most important ones have already been mentioned. I don't know a single respected author in the scientific community who still doubts that Darwin's ideas are fundamentally correct. All the discoveries since Darwin's day—think of the enormous number that have been made in genetics and molecular biology, fields totally unknown to Darwin and his contemporaries—have, one after another and without exception, confirmed his thinking.

That doesn't mean that all the questions raised by the phenomenon of evolution have been answered. The general scientific principle that the researcher's work is never done applies here too. The number of side issues and specific details still open to question is positively enormous.

Beyond this, even without any concrete evidence for it, we can't exclude a priori the possibility that the theory of evolution may one day be broadened or modified by the discovery of a new evolutionary factor, but such a factor could only be an objective principle, expressed in scientific terms. And this hypothetical discovery, while it might improve the theory of evolution as it is currently framed, could never disprove it. Thus Einstein's general theory of relativity further developed Newton's brilliant theory of gravitation without invalidating it.

None of the above statements is contradicted by the fact that all over the world scientists are concentrating all their efforts on challenging certain aspects of evolutionary theory, on looking for discrepancies and situations that may not be reconcilable with the theory in its current form. This is how science advances. According to the noted philosopher and theoretician of science Karl R. Popper, the habit of constantly calling its own findings into question and of trying to disprove them is the core of all scientific work. Popper's language may sound one-sided or exaggerated, but any scientist who gives up continual self-critical questioning of established knowledge will lose creativity and risk succumbing to ideological thinking.

Konrad Lorenz offers a paraphrase of the same principle with his remark that the most healthful setting-up exercise for a scientist is every morning after breakfast to take a favorite hypothesis and (tentatively) toss it out. The recommendation is surely, in principle, a good one, though scarcely practicable on a wide scale, for to stick with this exercise one would need an intellectual stockpile of Lorenzian proportions. But the point is clear.

Thus the ongoing research into evolution is in no sense an expression of lingering doubts about the basic correctness of what has been achieved so far. This situation is typical of all science; there is no absolute, final truth. Even the most solidly documented knowledge never goes beyond the status of a theory, of (at best) tried, tested, and broadly reliable theory.

To the scientist no theory can be trusted absolutely. None can ever acquire a dogmatic character that transcends all critical scrutiny. Every theory must always be examined in the light of new ideas and discoveries and, where necessary, be corrected. It's easy to see why: science can only progress by extending and overhauling previously elaborated findings and theories. This is the only way it can come closer to the truth about this world— a truth it will never grasp in its entirety.

And so the number of open questions, of problems that invite further patient investigation, is as large now as it ever was. Many people, fearful that a complete "victory" by scientific research might exclude all possibility of meaning outside the material plane, cling to this apparently consoling fact. Anyone who seriously thinks that the world will one day be reduced to a perfectly rational system, free of all mystery, has good reason for fear. And anyone who secretly suspects that, caught in the cold grip of science, the world will prove to be a deterministic trap, must be inclined to reject the findings of this grim science before he even knows what they are. The mere possibility of science's capturing the universe in an iron web of causality suggests the further possibility that concepts such as freedom of the will, moral responsibility, or even the meaningfulness of the world or one's own life could turn out to be mere illusions.

But we already have some initial reasons (others will come up later) that dispel this concern. The meaningfulness of life or the freedom of the will cannot be dismissed in advance as illusions (though neither can they be scientifically demonstrated). For some time now the real illusion has been the anxiety that such ideas can be refuted by science. But the world is not causally determined; Laplace's *daimon* has been dethroned once and for all and the nightmare he represented is over. It's just that the word, apparently, hasn't gotten around.

Viewed from the outside, then, an odd and paradoxical situation presents itself. Down through the centuries philosophy has discussed the possibility, among others, that the world should be interpreted as a causally determined machine. Philosophically speaking, the case could not be decided one way or the other, but the mechanistic model of reality had to be kept, so to speak, on the shelves of tradition.

It must be admitted, of course, that there was a time, several generations ago, when science largely identified itself with this model. But then, as the work of self-criticism went on, science was able to prove—and prove definitively, it would seem—that Laplace's *daimon* was a mere phantom. Nevertheless people are still afraid of him. And here is the paradoxical part—the reason they refuse to listen to science's arguments is that they're worried about something that those very arguments have long since shown to be groundless. Evidently scientists have to atone for the sins of their fathers to the third or fourth generation.

Under these slightly muddled circumstances a number of "religious" prophets have managed to win an audience. They try to persuade their listeners that there is a palpable limit, a sort of wall, at which the steamroller of scientific progress has to grind to a halt. As these prophets and their followers apparently see it, this limit neatly divides the world into two halves: one that can be successfully explained by science and the laws of nature, and the other where science will never be able to penetrate because natural law supposedly doesn't obtain there.

In the first half, from the standpoint of the search for meaning, science has, so to speak, left nothing behind but scorched earth. This is the mechanistic (or, as it's more often called, the materialistic) part of the world, the part that over the course of time and despite heavy resistance has had to be abandoned to the armies of science. But on the other side of the wall one can feel secure. The laws that govern that region will always be a closed book to science because they are beyond nature, are of the supernatural kind. And behind this argument lies the promise that the mere existence of the second half proves the existence not only of free will, of the meaning and purposefulness of creation, but even, in the final analysis, of God.

So much for vitalism and related points of view. We've already gone into the logical and theoretical flaws of this position; and we've noted that its idea of the scientific enterprise is about a hundred years behind the time. Still you can't, even today, simply shrug your shoulders and ignore this school of thought. It continues to stir a responsive chord among the laity, and its claim to be able to prove the existence of God offers a temptation to which, experience has shown, not even the churches are immune.[53]

For this reason I intend to bring together the most important evidence establishing the fact that this misguided effort to defend theology actually ruins it. By evidence I don't mean the reasons that lead to such ruinous conclusions—they're obvious enough. When the presuppositions of any ideology are as badly flawed as those of vitalism are, then all the inferences drawn from them, however carefully drawn, can only be intellectually rotten.

This is obvious enough and needs no proof. I wish to show

instead that vitalism, which strikes many people as attractive because they imagine that it supports and strengthens their religious faith, actually ought to be rejected on religious grounds. What follows then is not a scientific demonstration but an attempt at a critical scrutiny of the tenets of vitalism from a religious standpoint.

We have to begin with the limit, the "wall" we mentioned a while ago. Faith in its existence is the basic axiom of vitalism. On this point, at least, it's a step ahead of creationism, as it concedes the inanimate portion of the world to science. With the certainty that they have plenty of open territory (i.e., open questions) behind them, despite their constant need to retreat, vitalists never waver in their conviction that they are standing on the side of the fence that's ruled by supernatural laws and for that very reason unmistakably reveals its character of creation.

This is the pivotal point. Not the least part of vitalism's appeal comes from its implied proof for the existence of God. The cosmic terrain occupied by the vitalists bears, they think, manifest signs of having been created, which for most of them is concrete evidence of the Creator.

What religious objections can be made to this? First, the vitalists fail to see that while they hope to track God down in the animate portion of nature they are, by the very same method, expelling God from the inanimate world. Thus they themselves are actually doing what they once accused science of, namely arguing that God can be explained away—all the way out of the universe.

It's no good splitting hairs or beating around the bush here. Logic demands that if I assume that life is inexplicable and turn this into proof for the presence of God, then I must allow the use of explicability as a criterion for his absence. The more convinced I become of the Creator's existence in the face of certain unexplainable natural phenomena, the stronger my conviction must grow that phenomena I *can* explain function very well without him, or at least at a greater remove from him.

This is how the vitalist banishes the Creator of the universe from our everyday world into distant nooks and crannies; into the subatomic realm of nuclear particles, since their hybrid nature, part wave and part corpuscle, defies our powers of explanation; into the submicroscopic regions of cell nuclei, the site of the molecular coding mechanism, which Wilder Smith thinks cannot be accounted for in natural terms; or into the tubules of the kidneys, where concentration processes take place that, as Wolfgang Kuhn sees it, run counter to the laws of nature.

In an effort to be clear the foregoing may have been too strongly worded, but it's not false. If I claim to be especially close to God when I experience the (supposed) mysteriousness of certain phenomena, then I can't deny the obverse, that I must be especially far from him whenever I deal with comprehensible matters. In the light of this view the emergence of life appears to be a result and proof of divine creation, but the emergence of the solar system does not.

Vitalists miss the fact that the wall they so stubbornly insist on erecting brings about a twofold division. They divide the world not merely into a rational, empirical world and a supernatural one, but at the same time into one half that unmistakably presents itself to us as God's creative work and into another where this is evidently not the case, not in the same sense anyway.

Upon closer inspection the vitalists' world is revealed to be a cosmos all of whose parts are not equally deserving of admiration as God's creation. Or, putting it another way, the vitalists' world contains some phenomena that reveal themselves to be God's handiwork insofar as they require help from a transcendent source in order to exist or function, while it contains others that are, so to speak, not equally dependent on God because they can function on their own. There's no need of a long argument to show why such a distinction is religiously unacceptable. If God made the world, then he made the whole world with all its parts in the same divine fashion, without restrictions or qualitative differences.

But things get worse. The boundary line where vitalists take their stand is forever shifting. The historical growth of scientific knowledge forces that line to veer off ever more deeply into unexplored (unexplained) territory. The logical but truly grotesque conclusion here would be that the "sphere of influence" proper to the Creator and preserver of the universe is continually diminishing.

Grotesque, but unavoidable. Scientific progress goes on increasing the number of natural phenomena that can be explained in rational terms, that is, that don't need God to work. This increase on the whole may seem minuscule, but it's there. And so the Creator's power appears to depend upon the current state of research in, say, biochemistry or molecular biology or any other scientific field.

If that were so, then in Kant's day the presence of God could have been discerned in the solar system or in organic chemistry until Wöhler synthesized urea, whereas contemporary science has expelled God into the molecular microcosm. In this approach, which purports to be defending religion, God is actually "murdered inch by inch," as Anthony Flew, an English philosopher, aptly put it.[54]

That would be the case, if the presuppositions of vitalism made sense—fortunately they don't. We need not go over the reasons why they are logically and scientifically untenable. The only point to be established here is that the prophets from the vitalist camp who offer their services to the Church are false prophets, with whom the Church, in its own interest, should have nothing to do.

In between the area now open to human understanding and what is still *terra incognita* there lies a frontier that extends beyond the historical framework of some few millennia. In another sense this frontier has been receding for an incomparably longer time.

Right from the start there have been two dimensions of the unexplainable. First of all there are things that can in principle be understood but have yet to be discovered, such as the dark side of the moon (before space travel) or the genetic code (prior to the findings of Watson and Crick in 1953). But apart from that there is a level of reality that remains incomprehensible. A prime example of this might be the starry vaults of outer space, whose boundless finitude or finite boundlessness seems paradoxical to us—but not at all because of a lack of information (basically obtainable, only not yet on hand). Here we catch sight of a boundary beyond which our intelligence can't go, because behind this "wall" lie regions which the structures of human understanding can't deal with.

But even at *this* frontier the forces of vitalism can't make a confident stand, because it too is receding. It hasn't moved perceptibly during recorded history, but can we doubt that it has given ground since the days of Neanderthal or still earlier ancestors of our race? How could the realm of fundamentally incomprehensible things not be greater for Neanderthals than for us? And must we then assume that God was more present in that bygone world than he is today?

This argument also works the other way around vis-à-vis the future. Since we have no reason to believe that evolution after bringing forth humanity as we now know it has come to a halt, we must look upon ourselves as the Neanderthals of the future. Should we therefore believe that our descendants will, as compared with us, outgrow God's power and authority to the extent that their brains prove more efficient than ours?

Presuppositions that lead to such conclusions must have a fallacy in them somewhere. In this case I think it lies in a disguised variant of the old anthropocentric prejudice, which is innate and deeply rooted in all of us. So there's nothing we can do about it. But we ought to make use of our equally innate capacity for overcoming it, case by case, however painful that may be.

Blood had to be shed, literally, before a few critical thinkers finally convinced people that all the fixed stars in the sky did *not* revolve around them, that the earth was *not* the center of the universe. Our language won't let us describe events except in the form of statements about the doing or suffering of personal subjects, because this is the only way that suits our experience, where the whole world appears in a perspective flowing from the eye of the beholder.[55] As a result we even speak of a stone's liability to burst, without noticing the extent to which the animistic compo-

nent in the phrase falsifies the sober facts. And opposition to
Darwin's discoveries largely feeds on the widespread reluctance
to abandon faith in the differences separating us from all other
forms of earthly life.

Vitalism reflects this innate anthropocentric prejudice in a
veiled fashion. The boundary wall that it appeals to simply does
not exist in the objective world, being no more than the horizon
of our knowledge, advancing with that knowledge over the
course of our cultural and racial history. Anyone succumbing to
the lure of vitalism has simply been fooled by a projection of his
own mind, mistaking that for objective reality. That is to say, the
prime objection to this ideology is that it unreflectingly repeats
the error of making man the measure of all things; that it limits
the realm of an authority which it nonetheless views as "almighty" to the realm of the human IQ—in other words, that it
constitutes an attempt to cut God down to the size of man. Vitalists in fact are false prophets and false friends of the church.

11. Evolution as Creation

IT IS TIME for some preliminary stocktaking. In the previous chapters we sketched out the world picture that science has thus far pieced together. Naturally this could only be done in the crudest outline, touching simply on the aspects pertinent to our discussion. For this reason any readers with objections to this picture that have not come up for discussion should beware of assuming that their charges are irrefutable. If they look around for the answers with patience and good will, they will find them, for example in the books mentioned in the Notes. There's just no room in the text for exhaustive treatment of such problems.

Nevertheless, I insist that the scientific world picture traced out here is correct; it "adds up." But what does that mean? We have already seen that it can't be identical with the "truth" about the world. But neither can there be any doubt that it comes closer to this truth than all other sketches from earlier periods in the history of human thought.

And that says something of decisive importance for our argument. There is no going back from contemporary science's model of the universe. It is, to be sure, incomplete and provisional. It needs further development through continuous critical scrutiny, challenging its every detail. Future scientific progress will leave it further and further behind. But whatever happens, this model will never be completely displaced, never simply rejected *in toto* as "false."

Let's look at a typical example. Einstein's discovery of a connection between solid geometry and mass concentrations has provided us with a wholly new theory for understanding the nature of gravity: comparatively tiny planets revolve in orbits around comparatively enormous suns because the latter with their tremendous mass "deform" the space around them in such a way that the planetary orbits become the shortest paths a smaller body can take in such "curved" space. In other words, a planet revolves around its central star because its orbit corresponds under these circumstances to a "geodetic" line, that is, the line traced by a moving body when it's not deflected from its inertial course by an outside influence.

All that would be pure speculation, if this theory accomplished nothing more than Newton's classical theory of gravitation. It explains everything that Newton's theory can, but it also explains phenomena that were incomprehensible in Newtonian terms: certain peculiarities of Mercury's orbital motion (its so-called perihelion motion), the deflection of light in the sun's gravitational field (something Einstein was the first to discover, having predicted its occurrence), and various other things. Thus Einstein overtook Newton and advanced beyond Newtonian physics —but he didn't refute it. The formulas Newton applied to the movement of the planets are as valid now as they ever were; the computers governing the steering mechanism of space probes are programmed with their help.

The relationship, therefore, between Newton's theory and Einstein's is not one of false versus true. Einstein simply described the situation of a body in a gravitational field more precisely, in a way that allowed for a wider latitude of possibilities (e.g., extreme differences in mass, very high velocities), than Newton in his day managed to do. Einstein thereby came a little bit closer to cosmic truth than had any human being before him. But that didn't falsify Newton's explanation. In the retrospective light of modern findings Newtonian physics is merely less comprehensive, less broadly valid, than the modern variety.

A similar pattern holds in all other cases of scientific progress. The only total casualties in the wake of this advance are empty speculations and *ad hoc* hypotheses (of which science has naturally always had a limitless supply). But every theory that can be successfully used to explain some observable fact, however tiny, has demonstrated its hold on reality. No subsequent progress can do away with this.

Nowadays the mythical, animistic speculations of Stone Age people concerning the starry skies they watched each night are of interest at most to psychologists or students of prehistory. But the Neolithic observatory built at Stonehenge over three thousand years ago can still be used to determine the likelihood of a lunar eclipse. Evidently these ancient Britons had correctly recognized the periodic rotation of the so-called lunar node (the intersection between the orbits of the sun and moon), although of course they had not the slightest idea of what was actually causing this periodicity.[56]

We ourselves are only the Stone Age people of a future time that will leave what we now call modern science unimaginably far behind. So we have every reason to judge the current scientific world picture with the proper degree of modesty. Still we can safely assume that no matter how much progress science makes, it will never abolish or invalidate findings tested by observation and experiment.

This is why there can be no retreat from the contemporary scientific world model. Defective as it undoubtedly is, it nonetheless sets up a set of minimal conditions that our thinking must respect if it places any value on rationality. Given the immensity of the universe, our ignorance may be overwhelming, but we mustn't fly in the face of the little that we do know or thought will degenerate into superstition.

The same is true for religious language, for the formulas and images used by theologians to proclaim (that is, render comprehensible) their faith to both fellow believers and outsiders. This is not to attack metaphorical or mythological paraphrases of otherwise inexpressible matters.[57] But as soon as such paraphrases are meant or misunderstood to be literal truth, they convey nothing but pure superstition.

Since that is the case, theological language must submit to the criterion for meaningful statements established by the basic features of the modern scientific world picture. I don't have the slightest doubt that the crisis many church leaders feel is shaking their institutions derives in no small part from their reluctance to accept this requirement.

The criterion is a valid one, and so I consider the objections raised in the opening chapter (against two kinds of truth, etc.) to be solid. They arise perforce out of a framework established by the modern scientific model of reality. Thus the cogency of these objections depends upon the extent to which later chapters proved the (relative) validity of that model. This is why, for example, a critical review by the Church of the notion of humanity as "the crown of creation" is long overdue. Don't Christians realize how perplexing people find such ideas nowadays?

A cosmos that has finally been recognized as a historical process, a pattern of biological evolution that has been at work (on earth) for billions of years and has certainly not come to a screeching halt just now—against this background it is impossible to define modern man as the end product or supreme goal of all previous cosmic history. Anyone who tries to do so affronts reality in a way that increasingly strikes us as unthinkable.

Let it not be said that nobody makes such claims anymore. Currently, perhaps out of intuitive caution, religious representatives may not be making them as ingenuously and explicitly as they did a few decades ago. Still to this day the Christian understanding of man and nature continues to be marked by the conviction that the whole universe revolves around the human race and that *a fortiori* salvation history deals only with man to the exclusion of all other creatures.

How deep this conviction runs can be noted, for instance, in the writings of Teilhard de Chardin. Teilhard can never be credited enough for his attempt, truly revolutionary in its day, to integrate the fact of evolution into the structure of Christian faith. He went a long way in this direction, so far in fact that his church, which has no love for revolutionaries, kept the screws on him all his life. But unconventional as his thought was in many other areas, Teilhard still clung to one characteristic point of traditional dogma: man's unshakable position at the center of cosmic history. In *Man and the Universe* Teilhard argues that if the human race ever came to a premature end, the universe would fail its destiny. All hopes for future development, even of the cosmos, are bound up with the fate of humanity.

One hundred billion galaxies the size of our Milky Way have been observed (thus far!) in the universe, and for all that Teilhard insists that the fate of the cosmos depends on how things turn out on one planet, earth. This shows how irreversibly molded Teilhard was by his theological education, how hampered his thought was by "the folds of his priestly vestments," as Giordano Bruno regretfully noted with regard to his great intellectual model, Nicholas of Cusa.

Let it not be said either that such ideas are of no consequence, a remnant of the medieval world view that can be eliminated or reinterpreted with no ill effects. This anthropocentric fallacy, which swells to truly cosmic dimensions, could well have some very definite ill effects, as we shall see as we continue with the passage by Teilhard.

In the same chapter the author deals with the "final state of the earth" and discusses various possible shapes that future evolution might take. In this context he enumerates the dangers threatening such development, especially wars, revolutions, and other kinds of human failure. "The end might come in many ways," he declares, only to go on to say, reassuringly, that while all these perils are theoretically possible, they're actually out of the question, since we "have higher grounds for certainty, that *they will not take place*" (emphasis in the original). Then follows the explanation of these "higher grounds" in the form of the arguments already mentioned. Here Teilhard pronounces that humanity has something like a guarantee of survival on the sole basis of the supposedly irrefutable argument of man's central place in the cosmos.

In the world today tensions are building up like implacable natural forces between, on the one hand, various ideologies and cultures, and, on the other, the unchecked growth of the "nuclear club," the ever spreading capacity for atomic, chemical, and biological overkill. No one who looks out over this situation can be indifferent to the question of whether or not humanity will own up to responsibility for its own survival. Under the circumstances one is inclined to view faith in a transcendental, self-legitimating survival guarantee as positively dangerous, because it could contribute to further denial of responsibility.

I have no intention of imputing to any theologian, least of all Teilhard whom I personally admire, a wish to distract anyone from this responsibility. I simply want to bolster my contention that the wish to scrutinize the definition of humanity as the "crown of creation" relates to something more than just an academic theological controversy.

It's not easy to understand why people so obstinately put off making this long overdue correction of an error that goes back to a medieval vision of the world. Perhaps it's simply the result of a centuries' old habit, which in any case ought not to be underestimated. But I still fail to see how bidding farewell to an obviously outdated interpretation could endanger the substance of religion. It wouldn't even threaten our dignity, which can find solid foundations elsewhere. In all likelihood this new attitude would at least compel us to admit all other living creatures to a limited share of that dignity. I can see no harm in that.

Beyond this any further delay in coming to grips with science seems to run counter to interest of the Church itself. We can already point to the consequences of this delay, which are all the more to be regretted in that they might have been avoided. In 1972 the Evangelical Church let the Allensbach Institute circulate a questionnaire among its members, asking among other things what made faith difficult in this day and age. Of the answers suggested most people chose the one reading, "The scientific explanation of the world is completely different from Christianity's."

Can anyone doubt that the malaise expressed here derives to a large extent from the Church's unreadiness to break away from an outmoded medieval world picture? Anyone, however pious, who persists in clothing his message in a language belonging to a long vanished world ought not to be surprised if the suspicion arises that the content of the message might also be obsolete.

All of us can be moved by the language of the Psalms or an old hymn. But are we really going to let the matter rest there, with this emotional *frisson?* It's nothing more, after all, than a feeble reflection of the sensations that must have flooded over the men and women for whom these texts were not primarily a portion of a cherished, venerable tradition, but a living, up-to-date utterance.

Why don't theologians have the courage to blow on the embers that are smoldering under the ashes of an archaic language? That way they might awaken to a new life material which, though hidden beneath a thick layer of historical accumulations, still catches our attention by its glow. Why is it that all efforts in this direction always stir up nervous resistance? Why so much faint-heartedness from people who believe themselves in possession of a definitive, indisputable truth?[58]

And, in passing, let me raise another issue. Those who can't resist the temptation to win over simple souls by dressing up their transcendental message with graphic, highly concrete features inevitably make less simple souls suspect that the actual substance of their message could be made of the same flimsy stuff. I'm alluding to the descriptions—still widely encountered, especially in popular Catholic books and periodicals—that hold out to believing Christians the prospect of such things after death as a "glorified body," capable of levitation or telekinesis or other supernatural feats. This type of thing is painfully reminiscent of the fliers, full of wild promises, used for bait by certain occult sects when fishing among the people for converts.

There is no need to discuss the quality of such treatises. They may not be produced by the Church itself, but it's still unfortunate that they are tolerated and even given a quasi-legitimacy by the Church's imprimatur.[59] One would prefer to see it put such shoddy items on the Index instead of awarding them its stamp of approval, because as long as the impression remains that Christian faith requires the acceptance of such claims as true, it will put off men and women who are hoping for something more than superstitious come-ons.

Up to this point, however, the damage done by the Church's obstinate clinging to an antiquated world picture looks as if it can be repaired with relative ease. But this is not the case with the Christological problem. On this earth, which is integrated into cosmic history, where life has been undergoing a ceaseless evolutionary transformation for billions of years, the appearance of a single individual at a specific moment in this history cannot logically be taken as the supreme event, encompassing the entire process of development. Putting it more concretely, we cannot accept the identification of a son of God with humanity in its form during one particular stage of development as valid for all the countless other stages, past and future.

This identification must be viewed as an interpretive effort by a historical epoch which was convinced that human nature, in the shape of *Homo sapiens,* was fixed, final, and unchangeable. We have learned since then that this assumption is not true. In the face of evolution—a discovery now more than a century old—we can't help asking how far in time the concept of the "Incarnation" extends, given the limits of Jesus' historical personality. Does the "man" in "God-man" include Neanderthals and *Homo habilis?* And we have to pose the same question vis-à-vis the evolutionary future of our race, which might produce descendants for whom our "modern" notion of "human" might be as inappropriate as it is for the earlier hominids.

I am no theologian, and I'll not be so presumptuous as to speculate on the concrete effects this situation might have on the Church's image of Christ. Let me just make two more comments on this issue. First, the Church has to face the problem, unless it's willing to accept the continual lessening of its influence on people's minds, which (however vigorous the merely external, social customs of religion may be) has been noted on all sides by believers and unbelievers alike. And, second, the truth contained in the formula "God becomes man in the historical incarnation of Jesus" is not impugned by the attempt to free it from its linguistic wrappings. Nowadays these are beginning to distort that truth more than they preserve it, because they belong to a time frame that is no longer our own. In other words, the point is not to raise doubts about the Christian assertion that Jesus instituted a new relationship between God and humanity. The issue is simply whether it might not make sense to reflect on ways to translate this statement into language better suited to contemporary thought.

One final postscript: a warning should be issued against the temptation to dodge this problem by rhetorical pseudo-solutions or conceptual gimmickry. In Germany at least this tendency is most pronounced in the Evangelical Church. When God is defined as the "Cipher of Being," or religious truth as "a phenomenon existentially realized in personal encounters (with God)," which makes it fundamentally different from "other forms of truth" (such as scientific truth), then the theologian is spared any further conflict with the scientist. But that strategy of consistently avoiding conflict leaves persons in need of faith—those who would like to hear from theologians how they should envision God's activity in a world whose workings scientists have begun to explain to them with rational models—standing at the church door with empty hands.

In view of such modes of verbal evasion one is forced to concur with modern critics of religion such as Hans Albert, who speaks in this context of an "immunization strategy." This is chiefly calculated to empty the term "God" of all content so that there is nothing it can possibly clash with. As part of this arrangement theologians have to live with the fact that the distinction between faith in a "living God" and sophisticated atheism gets fuzzier and fuzzier.[60]

But enough scientific criticism of theologians. There's always a rather cheap smell to that sort of thing. Besides, anyone who voices such criticism must always expect to get the wrong kind of applause, from people who mistake it for an endorsement of their own atheistic or anticlerical position. This may give one pause, but my objections had to be raised here with complete frankness. Passing them over in silence would not only undermine the credibility of the reflections to follow, but would deprive them of the "space" they need to operate in.

But at this juncture I wish to argue that science doesn't simply place restrictions on what religion can say—it might also do just the opposite. Of course, people will never discover this if they keep giving science a wide berth for fear of endangering their faith. In the first place, this anxiety is self-contradictory. What value could a faith have whose strength, by the believer's own admission, is based on the systematic refusal to take cognizance of a certain kind of argument? Second, and above all, such fearfulness prevents the believer from encountering the aspects of modern science that *don't* put limits on religion but rather open up for it entirely new intellectual horizons.

This is what I wish to speak about—about examples, in other words, of scientific insights that bear positively on the religious interpretation of the world and of humanity; about the fact that between these two disciplines so long viewed as mutually exclusive it is possible to have a healthy relationship. This connection goes far beyond polite tolerance; indeed it sees science and religion as confirming and reinforcing each other.

I shall first sketch out the case for this in a purely speculative form. The scientific arguments that make the case plausible will then be presented in the second part of the book. This is the same method I followed in discussing the scientific arguments that establish a framework of conditions which religion must meet. Of all the positive consequences that modern science might have on the religious view of man and the world, the central one, I think, is the possibility of understanding evolution as the moment of creation—literally. I believe the idea deserves serious consideration: might not the process of cosmic and biological evolution (which strikes our imperfect brains as such a torturously long, drawn-out affair) be a creative instant?

A scientist would have no reason to object to the notion. He or she views "time" (inseparably connected to cosmic space) as something that came into existence some 13 billion years ago, together with energy, matter, and the laws of nature, in an event we've gotten used to calling the Big Bang. For a scientist, therefore, time is a characteristic of this world, along with energy, the plenum of space, and certain natural constants (the mass of elementary particles, gravitational constants, the speed of light, etc.).

Thus, in the modern scientific world picture, which so strangely transcends our naive ways of thinking, time is bound up with the existence of this world and cannot subsist without it. Time is not a comprehensive category, embracing the world as a whole, containing and determining it, as it were, from "outside."

If there is such an "Outside," it would be permissible to think of it as existing in a timeless state. I posit the reality of this "Outside" without any further justification, because unless one assumes a "Beyond" with respect to this world (some kind of transcendence), all talk of religious matters becomes senseless. But there is no reason why the temporal sequence of events in this world should necessarily be disjoined from that timeless Beyond.

It goes without saying that these statements are no longer scientific. From a strictly positivist standpoint they are neither true nor false, but simply meaningless. We shall see, however, that from other points of view it can seem quite meaningful to make them.

For the moment it should simply be noted that we are talking about statements that do *not* contradict any part of the modern scientific world picture. That in itself is no small thing. But, above and beyond that, they are only conceivable as a *result* of scientific work. Thus we have here the first instance of a scientific insight opening up a new path for religious interpretation.

Earlier epochs, naturally enough, tried to capture the mysteries of creation, transcendence, and their own ephemeral existence in the language and images familiar to them from their age and world view. Don't we have the same right? And don't we have to exercise it, unless we wish to see ourselves increasingly reduced to the role of museum attendants, reverently guarding the ideas of earlier generations (which carry us ever further back into the past) and eventually passing them down to others when we no longer understand them ourselves?

That is why I believe that evolution is identical to the moment of creation, that cosmic and biological evolution are the projections of the creation-event in our brains, and that the evolutionary history of both animate and inanimate nature is the form in which we witness creation "from within." "From outside," from the transcendental perspective, that is to say in actuality, creation is instantaneous.

Scientists will not gainsay this interpretation. More than that, they alone made it possible to begin with. Theologians ought to take an interest in it, because against the background of this idea answers spontaneously suggest themselves to questions that were unanswerable in terms of the older system.

Among these are, to start off with, the old problem of theodicy, of "justifying God's ways to man." How can we explain or, to put more pointedly and polemically, how can God be pardoned for the fact that he created a world that from the first has been full of every imaginable kind of suffering—pain and fear and sickness? How does evil enter the world, if it is God's creation? Ever since the days of Job every believer has had to wrestle with the problem of how to reconcile the imperfection of the world with God's almighty power.

But this contradiction becomes less acute as soon as we consider the possibility that the world as we know it might be an inchoate creation, as opposed to one that God had, so to speak, finished and sent off. It could be that the undeniable flaws and imperfections of the world are connected to the fact that it's part of a creation that is not yet complete. The believing person, for whom transcendence is real, can at least take consolation in the thought that this imperfection will turn out to be an illusion insofar as it is a temporally limited phenomenon and thus, in the light of transcendental truth, unreal.

If we take as our starting point the idea that evolution is nothing but creation in process (as we are capable of grasping it), then new and promising opportunities arise for broadening our comprehension of human existence. One such opportunity is the realization that we are evidently privileged to play an active part in bringing this creation to a climax, because ever since the human race awoke to consciousness, we have become increasingly responsible for the course of things in every part of the world that we have had access to.

From this we can deduce certain principles for human behavior that include all previous moral commandments—and on some important points actually complete them. This shows us that we are dealing with the sort of hypothesis that every scientist, and even theologians, might well take pleasure in. If human activity can have an effect on developments in the world, and if such developments are seen within the framework of a still unfinished creation, then that activity is subject, from the very start, to one supreme standard: it must at every moment be measured by whether it impedes or promotes the trend toward the completion of the world.

Does a given action, to take the simplest case, contribute to the lessening of pain, fear, and sickness, or not? But also, does it, precisely on account of the basic moral meaning of the connection between individual behavior and the fate of the self-completing creation, increase the chances for other men and women to recognize and learn about that connection? This second imperative leads to a whole spectrum of concrete rules, including such things as fighting illiteracy and hunger, both of which in different ways prevent people from seeing their part in the process of creation.

Up to this point the only ethical demands mentioned were ones that were already long established, though on different grounds (and though it may seem that this way of justifying them is more concrete and more persuasive to a contemporary mind). Two further moral consequences of this approach strike me as still more important. They relate to religious imperatives that, as historical experience shows, have not always enjoyed the same self-evident validity.

Thus, Christians were always a little in danger of seeing concern for their own salvation as the legitimate core of their moral obligations. True, the Pharisee was forbidden to imagine himself superior to the publican. In addition, he had to "do good" to the latter and was even required to love such people "as himself." But all these commandments in no way barred the thought that the other's final damnation would, after all, leave one's own chances for redemption unimpaired.

This view changes radically as soon as you look upon the world we live in as creation-in-progress. One's hopes can no longer concentrate on the possibility of an individual, totally isolated redemption, that is, of being transported after (or by means of) death out of a world of imperfection into the perfection of a transcendent other world. Redemption in the fuller sense can be conceived of only as taking part in the salvation of the world as a whole, in the salvation of the entire cosmos at the moment when its history has concluded and the instant of creation has passed.

But until then the decisive factor in the course the world takes (insofar as human activity helps to decide it) is quite obviously not the individual. It is the behavior of the entire group that decides what detours we force ourselves to take on our way through history. Realizing this might make it easier to avoid the mistaken notion that the "good" one must do unto others serves primarily as evidence that one deserves salvation. Instead an action might be considered good only if and to the extent that it contributed to the building up of a moral solidarity that included everyone without exception, since everyone has some influence on the course of history—that history where everything hangs in the balance for all of us.

But, in my opinion, a still more important moral consequence of this broader solidarity arises in connection with the relationship between the believing Christian and this imperfect, incomplete world around us. It relates to a certain kind of peculiarly Christian denial of this world and its imperfections—I mean the tendency to what Ernst Bloch mockingly called "Christian drivel about the other world," the sort of thing that Marxists have always indignantly denounced and not without some justification. (Self-critical modern theologians suggest that this tendency spurred the rise of Marxism in the first place.)

The issue here is the "theory of the fundamental otherworldliness of man, of human destiny and intentions," which has again and again misled Christians into "directing their actions not to the betterment of this world, but to redemption from it." The author just cited, Rupert Lay, speaks of a theory that conceals a basic misunderstanding—the error of seeing a spatial relationship between this world and the next. He argues that this misunderstanding seduces people into a strategy of trying to overcome this world and renounce it. This has brought with it the danger of a breakdown on social and humanitarian issues by the Christian churches and turned "fleeing the world" into a Christian byword.[61]

Lay, who is a well-known Catholic theologian, criticizes the notion that envisages this world and the "Beyond" as two adjacent regions. One is the realm of untrammeled godlessness, imperfection, and pain; the other is imagined as a perfect paradise, free of all suffering. From here it's not far to the notion that exhaustive efforts on behalf of this imperfect world are wasted anyway, that what matters is concentrating one's attention as soon as possible on that other, paradisiacal world.

When this world and the other are thus separated from and opposed to each other as if they were alternatives, the temptation inevitably arises to steal away like a deserter from the imperfect, pain-filled world. This is one of the basic reasons why the Christian churches have all too long closed their eyes to social questions and confined themselves to "putting flowers on the chains of the oppressed, instead of smashing them."[62] Has this decorative approach altogether disappeared today?

But identifying evolution and the "history of creation" makes that Christian "two-tiered universe" impossible. It precludes thinking of the world as a place frozen in imperfection, whose shortcomings can be avoided only by running away. It now turns out that all of us, like it or not, constantly participate in changing the world over the evolutionary course that will bring it to its consummation. No one, therefore, can escape the responsibility arising from the fact that what one does and, just as important, doesn't do, within the limits of individual potential, will play a part in determining the pattern of development that decides the fate of the cosmos.

Let me content myself with those few suggestions. An outsider can't dictate to theologians how they are to interpret the world and humanity's place in it, or how to describe the peculiar religious content of the relationship between humanity and world. The examples I have given are designed simply to show that, contrary to popular opinion, science's view of the world in no way necessarily contradicts the statements religion makes about that same world. Science does limit the possibilities of religious discourse within the framework of certain minimal conditions governing logic and matters of fact. These could be seen as a sort of scientific "prolegomena," a set of preconditions that theology has to pay heed to, if its statements are to be taken seriously in the context of contemporary intellectual culture.

My examples should prove that modern science provides theology with concepts and images of the world that open up brand new avenues to religion. The attempt to travel these new roads seems to me a legitimate one, for these scientific concepts and images have become the accepted language of our age exactly as did those used by earlier periods, according to their best lights, to grasp and describe the same religious insights.

But for the time being all these examples offer us nothing more than assertions pure and simple. The conclusions drawn from them may be plausible and internally consistent. But this does not spare us the obligation to adduce reasons to justify them. Yet doesn't science, which is basically positivistic and oriented toward objective knowledge, run counter to this whole discussion on one critical point—doesn't it reject as unverifiable any idea of a reality outside our world, of transcendence, of a Beyond?

Doesn't it maintain that even asking questions about such a possibility is meaningless?

"If there is an objective truth (as the materialists think), if only science (by reproducing the external world in human experience) is capable of mediating the objective truth to us, then any kind of fideism must be unconditionally rejected."[63] Is that actually the last word? In point of fact, can't the possibility of fideism, of believing in a reality beyond our world, be refuted and laid to rest? Where could there be room for such a reality, considering our daily experience of the objective, undeniable reality of the world around us? Where indeed, considering that science owes all its successes to the systematic way it dispenses with all hypotheses that cannot be objectified, verified, or falsified?

This is an objection that has to be faced. We must therefore look into the question of how real our "reality" actually is, and for that purpose we shall have to go off on another scientific digression.

NOTES

1. One of the laudable exceptions is the book *Does God Exist?* by Hans Küng (Garden City, N.Y.: Doubleday, 1980), to whom I am indebted for many stimulating suggestions.
2. See Sigrid Hunke, *Glaube und Wissen* (Düsseldorf: 1979).
3. Hans Albert, a militant atheist and student of Popper's, gives some impressive examples in his *Traktat über kritische Vernunft* (Tübingen: 1975). See the following note as well.
4. The doctrine of the "two truths" can be pressed to rarefied heights of abstraction, as shown by the work of certain Protestant theologians. Thus Emil Brunner defines religious (as opposed to scientific) truth as "a phenomenon existentially actualized in personal encounter" (quoted in Jürgen Hübner, *Theologie und biologische Entwicklungslehre* [Munich: 1966], p. 245). In a similar vein theologian Fritz Buri describes God as "the mythological expression for the unconditional nature of personal responsibility" (quoted in Küng, *Does God Exist?* p. 335).
5. A longitudinal study released in 1981 by the Allensbach Demographical Institute documents the rise of alienation from the Church. In 1953 among German Catholics 55% of the men and 64% of the women still attended church regularly. By 1979 the figures were only 27% and 44% respectively. In 1953 44% of all Catholics still supported the indissolubility of marriage; today it's

only 12%. Among Protestants the number of regular churchgoers sank over the same period from 18% to 9%. In 1953 8% of Catholics and 13% of Protestants *never* went to church. Nowadays the figures are 13% and 21%.

6. Extraordinary as Einstein's achievement was, it was naturally not the only decisive factor that transformed science's view of itself. (This revolution had been in the making at least since Immanuel Kant.) But the theory of relativity makes it especially clear why and in what sense scientific progress compelled us to realize that our intelligence is inadequate to grasp the reality of the world.

7. "The last proceeding of reason is to recognize that there is an infinity of things which are beyond it. It is but feeble if it does not see so far as to know this. . . ." Blaise Pascal, *Pensées* (New York: Dutton, 1958), no. 267.

8. On this point Pope John Paul II gave an encouraging address in the cathedral of Cologne on the occasion of his visit to Germany in 1980. The pontiff insisted that there was in principle no conflict between reason and faith, and said that the Church regretted its past intervention in the scientific quest for knowledge.

9. "We are not deluding ourselves. Atheism today demands an account of our belief in God as it never did in the past. In the course of modern times, this belief has been increasingly on the defensive and today has often been silenced, at first with a few people and then with more and more. Atheism as a mass phenomenon, however, is a phenomenon of the most recent times, of our own times. The question is forced upon us: How did it get so far? What are the causes? Where did the crisis break out?" Küng, *Does God Exist?* p. xxii.

10. See Part III, note 32.

11. Even Teilhard de Chardin, remarkably enough, did not draw this conclusion. In *Man in the Universe* he makes the apodictic statement that "He [man] could never come to a premature end or a complete halt nor could he go to ruin, without the universe simultaneously failing its destiny." A few lines later Teilhard announces laconically that "Man is irreplaceable." In reading such passages one has to be fair to Teilhard and recall the intellectual isolation his religious superiors kept him in. As early as 1926 he was deprived of his teaching post in Paris, he was not allowed to accept offers at universities elsewhere, and he never saw any of his major works published in his lifetime.

12. The Doppler effect explains the flight of all the galaxies in the universe away from each other through the "red shift" (i.e., of the spectral lines of light emitted by these systems), which increases with distance. The theory of the Big Bang was arrived at by taking the various rates of speed of this galactic flight and extrapolating back to the beginning of the world. All this has been written about so much in the last few years that I assume readers are familiar with it.

13. I have borrowed this example from Paul Davies' *The Runaway Universe* (London: 1978), pp. 39ff. See also Steven Weinberg's *The First Three Minutes* (New York: Basic Books, 1976).

14. For particulars consult my *Im Anfang war der Wasserstoff*, 5th ed. (Munich: 1979).

15. See "The Science-Textbook Controversies," *Scientific American* (April 1976), pp. 33 ff.

16. Strictly speaking, this approach is objectively correct, but it's nonetheless

one-sided, too skewed by the perspective of individual (human) learning capacity and measured too exclusively against that norm. It is precisely this "stupid" unteachableness that is the reason for nature's quite unbelievable flexibility and adaptability, which becomes apparent every time the environment changes. Nature's repeated deviations can be regarded as "mistakes" only so long as the norm corresponds to optimal adaptation to the existing environment. In the final analysis snowshoe rabbits, ptarmigan, and polar bears blend in so well with their Arctic surroundings because of evolution's "inability" to give up experimenting with abnormal pigment variations even after all its past failures.

17. See Margaret O. Dayhoff, "Computer Analysis of Protein Evolution," *Scientific American* (July 1969), pp. 86 ff. and Francisco J. Ayala, "The Mechanism of Evolution," *Scientific American* (Sept. 1978), pp. 48 ff.

18. The point argued here with the example of cytochrome C is also valid for numerous other enzymes, even though their pattern of variation has not been traced in as many species.

19. This is simply because the common gene pool represents the definitive criterion of a biological species. There can be no new species, as far as a biologist is concerned, until its members are incapable of interbreeding with those of the original species. As a rule this is the result of differences that are either morphological (inability to copulate) or physiological-genetic (e.g., in the form of a "sterility barrier," due to immunological incompatibility between the gametes of the two species).

20. We might reflect that the only reason why this grain was able to become our "daily bread" is that it consists of a substance related to our own bodies. For a substance to be suitable for food requires among other things that its basic elements be identical to those ɔf the organism wishing to feed on it.

21. There are, of course, many other unrelated arguments for the reality of evolution. I especially recommend Günther Osche's *Evolution*, 6th ed. (Freiburg: 1975), pp. 11–30.

22. Immanuel Kant, *Allgemeine Naturgeschichte und Theorie des Himmels* (Leipzig: Reclam, n.d.), p. 178.

23. Cited from memory. I have been unable to find the original passage.

24. Jacques Monod, *Chance and Necessity* (New York: Knopf, 1971), pp. 172–173: "Now does he [man] realize that, like a gypsy, he lives on the boundary of an alien world? A world that is deaf to his music, just as indifferent to his hopes as it is to his suffering or his crimes." Or page 180: "The ancient covenant is in pieces; man knows at last that he is alone in the universe's unfeeling immensity, out of which he emerged only by chance."

25. The term "entropy" was introduced into thermodynamics in 1865 by Rudolf Clausius. To the physicist entropy means the degree of order (= improbability of its being in a given state) present in a "closed system." For example, a metal rod heated on one end incorporates a higher degree of order (= lower probability of such a state = less entropy) than a rod of uniform temperature. In all systems—so long as there are no outside influences—entropy is always increasing, never decreasing: the temperature difference in the metal staff heated only on one end will always even out, will never reestablish itself. The same is true of all differences in energy, even for the entire universe (assuming that it is a closed system, i.e., finite). The fact of entropy makes it possible to distinguish physically the past from the future. The past

of any system will display a lower level of entropy. Living organisms, by the way, are not closed systems because they depend upon an external source of energy (food). Nevertheless, as a whole they inevitably increase the sum of entropy and cannot escape the consequences of this fundamental principle of nature for an unlimited period of time.

26. Fred Hoyle and Chandra Wickramasinghe, *Lifecloud* (London: 1978).

27. *Im Anfang war der Wasserstoff*, 1st ed. (Hamburg: 1972), pp. 126 ff.

28. "More Evidence for Cometary Lifecloud," *New Scientist* (Feb. 28, 1980), p. 655.

29. Hoyle and Wickramasinghe, *Lifecloud.*

30. Monod, *Chance and Necessity.*

31. It should be noted that we are are speaking here of basic research, not of applied science, much less technology. Nowadays discussions about the problems and risks of "scientific progress" often blur the differences separating these activities.

32. For a comprehensive treatment of the emergence of life see the brilliant book by Manfred Eigen and Ruthild Winkler, *Das Spiel* (Munich: 1975); in English translation, *Laws of the Game: How the Principles of Nature Govern Chance* (New York: Knopf, 1980).

33. To take one example, Wilder Smith's *Die Naturwissenschaften kennen keine Evolution* (Basel/Stuttgart: 1978) is full of such howlers as his remark on pp. 64–65 that he "rejects" the scientific (statistically defined) concept of information. This is roughly equivalent to a modern physicist rejecting quantum mechanics.

34. Yehoshua Bar-Hillel, "Wesen und Bedeutung der Informationtheorie," in *Information über Information* (Hamburg: 1969); Karl Steinbuch, *Automat und Mensch*, 4th ed. (Heidelberg: 1971).

35. Wolfgang Kuhn took his Ph.D. in Geography (!). In 1962 he was appointed instructor in Biology at the Pädagogische Hochschule (School of Education) of the Saarland at Saarbrücken. In 1978 Kuhn was transferred to the Faculty of Science and Mathematics of Saarland University with the rank of university professor—a strictly political decision. Members of the Biology Department protested against Kuhn's appointment, but they were overridden, while their colleagues in Zoology posted a public announcement dissociating themselves from Kuhn's ideas.

36. Eigen himself occasionally speaks of "molecular semantics" in describing how the genetic code functions (*Das Spiel*, pp. 304 ff.), but he means this in the purely metaphorical sense. Eigen does *not* draw any analogies to semantics in the true linguistic sense as a way of explaining the development and function of the genetic code. Similarly C.F. von Weizsäcker writes, "Chromosome and growing individual are related to one another *as if* the chromosome were speaking and the individual listening" (*Die Einheit der Natur* [Munich: 1971], p. 54, emphases in original); in English translation, *The Unity of Nature* (New York: Farrar Straus Giroux, 1981).

37. Wilder Smith, *Die Naturwissenschaften kennen keine Evolution*, p. 30.

38. Eigen and Winkler, *Das Spiel*, pp. 142 ff.

39. All living creatures now found on earth are doubtless "monophyletic." But the fact that they all stem from "one root" in no way rules out the possibility that life may have made several, perhaps even countless, other attempts at getting started on earth.

40. In Darwin's day the fact that species underwent change had long been recognized by the vast majority of scientists. Darwin's achievement lay in his explanation of the causes of this change. Despite the relatively narrow empirical basis that he had to build on (the mid-nineteenth century knew nothing of the genetic code, molecular biology, etc.), not a single one of all the subsequent discoveries by biologists has contradicted his theory of evolution.

41. See Günther Osche, *Die Evolution,* 6th ed. (Freiburg: 1975); Heinrich K. Erben, *Die Entwicklung der Lebewesen* (Munich: 1975); Rupert Riedl, *Die Strategie der Genesis* (Munich: 1976).

42. Riedl, *Die Strategie der Genesis,* p. 317.

43. Ilya Prigogine, *Vom Sein zum Werden* (Munich: 1979); in English translation, *From Being to Becoming* (San Francisco: W. H. Freeman, 1980). Prigogine's book is highly abstract and crammed with mathematical formulas. Let me try to outline, by way of a simplified example, one of the critical points of his extraordinarily important work. Among other things, Prigogine notes that the idea of the physical world's absolute determinacy is acknowledged today as a result of "excessive idealization" in classical mechanics. What does that mean? Consider the laws of falling bodies, discovered by Galileo. Galileo himself was perfectly aware that his mathematical description rendered the findings of his extensive experiments in an "idealized" fashion. In reality the behavior of the various balls and weights he used always deviated slightly from the "law," owing to "disturbances" from friction (on his inclined plane) or wind resistance. What Galileo and all later physicists meant when they said that the law of falling bodies "determines" the behavior of such bodies, was, strictly speaking, only this: a falling body *would* behave as the law said it was supposed to, *if* it fell under ideal (disturbance-free) conditions. But there is no truly ideal place for experiments anywhere in the universe. There will always be gravitational attraction from other heavenly bodies, there will never be an absolute vacuum, and particle radiation will always have some influence on the outcome of the experiment. All of these "disturbances" have to be deducted from the results before one can say that the case in question conforms to the law.

Only recently have physicists begun to suspect that their description of the world may have been only too idealized, insofar as they eliminated (as "disturbances") real outside effects from their mathematical models. Prigogine has closely pursued this matter and in the process made some revolutionary discoveries. He has, for instance, worked out a "physics of becoming" by showing that, contrary to the assumption of classical physics, the laws of thermodynamics do not always lead to the disintegration of ordered structures, but under certain circumstances they can also bring about their spontaneous development, even in the macroscopic realm.

44. Erich Thenius, *Lebende Fossilien* (Stuttgart: 1965).

45. This is unquestionably an anthropomorphic way of speaking. Later on in the text I shall attempt to validate my use of such expressions. For the time being let me simply cite the remark of Immanuel Kant, "Hence we are . . . quite justified in speaking of the wisdom, the frugality, the providence, and the beneficence of nature, without thereby making her into a rational creature" (*Critique of Judgment,* § 68).

46. This doesn't mean, as one occasionally hears people say, that either Clausius (who developed the concept of entropy) was right and Darwin wrong or

vice versa. See Eigen and Winkler, *Das Spiel,* pp. 116 ff.

47. Protozoa, and especially bacteria, are the favorite subjects for experiments with evolution, because these species reproduce so quickly (two or three generations per hour in the case of bacteria) and in such immense numbers, which leads to a proportionately high absolute rate of mutation. Since the genetic code is the same in almost all living beings, the results obtained are applicable to all other species, as far as the basic laws of evolution are concerned. I am indebted to Günther Osche for telling me about the experiment described in the text.

48. Konrad Lorenz, *Die Rückseite des Spiegels* (Munich: 1973); in English translation, *Behind the Mirror,* trans. Ronald Taylor (New York: Harcourt Brace Jovanovich, 1978).

49. Konrad Lorenz, "Über die Wahrheit der Abstammungslehre," *Naturwissenschaft und Medizin* 1 (Mannheim, 1964), p. 5.

50. Nietzsche interpreted Darwin's phrase to mean this sort of *bellum omnium contra omnes* ("war of all against all"), and in the same passage he explicitly speaks of the supposed "privilege of the stronger." (*Thoughts out of Season,* quoted in Küng, *Does God Exist?* p. 351.) This is significant because such statements led extreme Social Darwinists to make Nietzsche their star witness.

51. Osche, *Evolution,* pp. 11–30.

52. Ingo Rechenberg, *Evolutionsstrategie: Optimierung technischer Systeme nach Prinzipien der biologischen Evolution* (Stuttgart: 1973).

53. Thus Wilder Smith's books may be found circulating in Catholic seminaries —and creating a good deal of confusion, as I know from personal experience. And Wolfgang Kuhn (see note 35) is more or less regularly invited to official Catholic functions, including liturgical services, to present his ideas.

54. Anthony Flew, quoted in Küng, *Does God Exist?* p. 332.

55. This "subject-centrism" (R. Bilz, *Befinden und Verhalten,* ed. J.D. Achelis and H. von Ditfurth [Stuttgart: 1961], p. 161) is a clear reference to the criteria for natural selection used by evolution in producing the human brain. Evidently the mutations selected for were not those that would have improved human performance in grasping and assimilating the objective features of reality, but those that provided the brain's possessor with better chances for survival.

56. Fred Hoyle, *On Stonehenge* (Oxford: 1972).

57. See Leszek Kolakowski, *Die Gegenwärtigkeit des Mythos* (Munich: 1974).

58. There are exceptions: some theologians *have* taken up the challenge. The attempt to "demythologize" Christianity is associated above all with the name of Rudolf Bultmann. See the debate between him and Karl Jaspers in *Die Frage der Entmythologisierung* (Munich: 1954).

59. For example, Wilhelm Schamoni, *Theologisches zum biologischen Weltbild* (Paderborn: 1964), pp. 90 ff.

60. Albert, *Traktat über kritische Vernunft,* pp. 115ff.

61. Rupert Lay, *Zukunft ohne Religion?* 2d ed. (Olten: 1974), pp. 42 ff. The author is a Jesuit and holds the chair for Philosophy and Scientific Theory at St. George's Catholic University near Frankfurt.

62. Quoted in Küng, *Does God Exist?* p. 237.

63. Thus says Lenin in *Materialism and Empiriocriticism.* Quoted in Küng, *Does God Exist?* p. 240.

PART II

Objective Reality and the
World Beyond

12. How Real Is Reality?

FOR THE NAIVE realists the possibility of anything "beyond" the reality of our world is automatically confuted by referring to their immediate experience. Those who will believe only what they can "get their hands on"—the things whose reality they think they can ascertain by being able to see them, hear them, or in some other way perceive them—must treat all talk of such a possibility as chimerical. But the objective basis that such realists, unshakably confident about perceptible facts, appeal to is far less solid than they think. For more than two thousand years, ever since Plato, naive realism has been exposed as an illusion. What are the arguments in the case?

A few years ago someone asked me whether it would actually get dark in the universe if all eyes disappeared. Questions of this sort stand at the door of every epistemology. "Dark" and "bright," as anyone who takes the trouble to think about it can tell, are not characteristic qualities of the world but "vision experiences," that is, perceptions that arise when electromagnetic waves of a certain length—between 400 and 700 millionths of a millimeter—fall on the retina. We have every reason for assuming that this is true of animal eyes as well; and we even know that the length of the waves which create the impression of brightness differs in many animals from the frequencies that do the same for human eyes.

Needless to say, for the impression of sight to arise it is not enough that the fundus (back part) of the eye should be struck by waves of just the right length for retinal cells to respond to.

There is a further prerequisite: the waves must be transmitted from the retina to the brain, to a quite specific region of the cerebral cortex around the occiput, the so-called visual cortex. As far as we know, the electrical and chemical processes that take place in this layer of neurons (only a few millimeters thick) are the terminal point of the physical operations that underlie our optical experiences.

For several decades now scientists have been using ingenious methods to investigate what goes on when the lens of the eye projects an image of the exterior world on the retina, which the latter breaks down into an immense number of nerve impulses and then passes on to the optic nerve. The findings of this research have been quite exciting, but we are as much in the dark as ever with regard to the crucial processes of sight. It is literally dark, by the way, in the visual cortex, even when we see something. There is also nothing like a picture in the brain. (Who could look at it there?) The way in which the visual cortex deals with the electrical impulses transmitted to it by the optic nerve has not the slightest resemblance to any copying mechanism we know of.[1] Even the connection that exists—and must exist, there being a demonstrable dependency here—between these electrochemical events and the experience of seeing remains absolutely mysterious.

So it's not "light" anywhere on the path leading from the retina to the visual cortex, not even at the end point. It only becomes "light" in the experience of seeing, which lies beyond that ever puzzling frontier separating (as far as our mind is concerned) physical events from psychic experiences. Thus it's not "light" in the external world, in the cosmos, regardless of whether there are eyes or not.

Does that mean the universe is actually dark? The original question had presumed this might be the case, but that possibility is also out. The adjective "dark" does not refer to a characteristic of the outside world either, but likewise describes what is exclusively a visual experience. One could also say that since the cosmos can't be light, it also can't be dark, for one is conceivable only as the contrary of the other.

The only way to do justice to the situation, then, is to accept the fact that in the external world there are electromagnetic waves of the most widely varying lengths (or frequencies), that our eyes are susceptible to a (comparatively very narrow) section of this frequency band, and that our brain, or more precisely the visual cortex, in some absolutely mysterious fashion, translates the signals emitted by the retina into optical experiences, which we describe with the words "light," "dark," and the various terms for color.

So the apparently simple question of whether it would be dark in the world if there were no eyes proves to be a major one. And incidentally our discussion of it has thrown us up against all the basic problems of epistemology. We have, first of all, assumed that outside our experience an external world really does exist. Second, we found that we cannot simply regard what we experience as a real quality of that external world. And finally it has already been shown that in all likelihood there are real qualities of this world which we are quite incapable of perceiving, for example, the electromagnetic waves whose frequencies lie beyond the narrow perceptual range of our retina.

Over the last hundred years scientific and technical progress has made it possible for us to detect, at least indirectly, other parts of this band of frequencies and even to use them for practical purposes (e.g., X-rays and radio waves). Technology has provided us with artificial sense organs, and with them we can prove that the features of the external world exceed our capacity to perceive them. This fact suggests that there are probably an unimaginable number of other objective qualities in the world about which we shall never learn a thing, not even indirectly. It would surely be very strange if the outside world did *not* contain things beyond the reach of our instruments.

And, as if all this weren't enough, even the tiny slice of the outside world that we *can* grasp is in no way mediated to us by our brain and sense organs "just as it is." In no instance is what finally emerges in our experience a faithful copy of what's out there. Even the little that we finally do perceive doesn't get to our consciousness without first being processed in a complex and, as far as the details go, utterly obscure manner. Our senses do not give us a copy of the world—they interpret it for us. The difference is fundamental.

A few examples should suffice to convince us. We have already noted the fact that our eyes and brain transform electromagnetic waves into the experience of light and create various color impressions, depending upon the length of the wave striking the fundus. But the nature of an electromagnetic wave has nothing whatsoever to do with what we call "light" or "bright." (With the one crucial exception that one constitutes the cause of the other, the moment that eye and brain are involved together.) Brightness and electromagnetic waves bear no resemblance to each other at all.

The same is true for the different colors. A wavelength of 700 millionths of a millimeter has just as little to do with the sensation "red" as a wavelength of 400 millionths of a millimeter has to do with the sensation "blue." And there is no similarity between, on the one hand, the physical distance separating the two frequencies (only 300 millionths of a millimeter) and, on the other, the psychic contrast separating the colors red and blue.

One last example. It was just stated that we cannot immediately perceive electromagnetic waves outside the narrow band of optically visible light. Strictly speaking, that is not quite true, but the exception to this rule makes the whole business all the more bewildering. At a different place on the same spectrum (roughly between wavelengths of a thousandth and an entire millimeter) we can register other waves. Only it's not our eyes that respond to them, but the sensory receptors in our skin. We don't see these waves, we feel them. We perceive them as thermal radiation.

The reader must realize what this means. All electromagnetic waves are identical in character, always displaying the exact same kind of radiation. The only difference consists in the wavelength. Depending upon the specific adaptation of our sensory cells we experience certain frequencies of these waves as light or as various colors—or as radiant heat. So quite obviously there can be no talk of "copying" the real world "just as it is."

It's clear that naive realists are indeed naive. In no time at all their seemingly rock-solid notion of a reality that can be verified (objectified) through sense perception crumbles into pure illusion. Things are not so simple. But philosophers didn't need to wait till the discovery of the physiological facts of perception mentioned here. Naive realism can be shown to be untenable in an even more rigorous and logically more forceful manner, with abstract, purely philosophical arguments—though not, of course, without some reference to human sense perception. For more than two thousand years, in fact, it has been clear that of all the world surrounding us only a small section is accessible, and that this fragment is mediated to us by our senses in a mode not just highly imperfect but positively distorted.

In the fourth century B.C. Plato saw through the situation we find ourselves in vis-à-vis the outside world and described it in a famous allegory. The situation of humanity, he said, resembles that of prisoners chained in a cave with their backs to the entrance. The only thing they can see are shadows projected on the wall in front of them (by a fire behind them illuminating models of things from the world outside the cave). But the prisoners take the shadows for reality, and so they are doubly deceived. Thus the philosopher's noblest task is to enlighten people about their true situation. They must at least realize that they wouldn't have the real world before them until they were able to turn around and look out of the entrance to the cave, or if they could leave the cave and see the things themselves, instead of relying on the flickering shadows of their images on the wall.

To this day no one has summed up the human condition in a more sublime or telling analogy. Equally powerful is the admonition that as long as people take at face value what their senses tell them about the world (as along as they take the shadows of the world for the world itself), they remain, intellectually speaking, children. Thus ever since Plato there has been a special philosophical discipline called epistemology (or theory of knowledge), which is exclusively concerned with discovering how our knowledge and experience of the world actually work.

Two and a half millennia since Plato's day this question has still not gotten a definitive answer. Recently, however, as a result of a revolutionary fusion of philosophical (epistemological) and scientific ideas (once again breaking down a barrier that up till then had been deemed permanent), the problem of human knowledge seems to have entered a crucial and possibly final stage of investigation. Obviously we have to limit our attention to the stages of this process that are most relevant to our purposes. We shall concentrate on two: the first is bound up with the name of Immanuel Kant; and the second is the "evolutionary" theory of knowledge developed by Konrad Lorenz, Karl Popper, and a number of others over the last few decades.

13. Reality Has No Handles

"READING Kant on the side is totally impossible," Konrad Lorenz declared in an interview not long ago. By his own account the Nobel prize winner and student of animal behavior has read only one work by the Königsberg philosopher in his entire life: the *Prolegomena to Any Future Metaphysics* (a sort of abridged version of Kantian epistemology). The number of those who have come to the same conclusion (namely that Kant is unintelligible except to professional philosophers who are also Kant specialists) is larger than many think. It's just that most people don't have the nerve to admit it as unabashedly as did the grand old man of ethology. For this reason Lorenz's public confession has a doubly liberating effect—even if he was deliberately and provocatively exaggerating (it wouldn't be the first time).[2]

The statement might have a provocative effect on a goodly number of his scientific colleagues (Lorenz likes to amuse himself by doing such things) and with reason: the man who announced that Kant was too obscure for him has given the latest turn, and a perhaps decisive turn at that, to the theory of knowledge. And in so doing he has provided Kant's philosophy with a new foundation and at the same time gone beyond it.

This is consoling in two ways: any members of the anonymous confraternity of those who have been vanquished by the texts of the "Königsberg Chinaman" (to quote Nietzsche) will be relieved to know they're in good company. But Lorenz's example shows above all that it is possible to grasp the core of Kant's insights even if one cannot penetrate the works where Kant expounds them.

To this end let us return to our naive realists, whom we last saw amidst the shambles of their realistic world view. What do they have left? For all the skepticism that has since suffused their minds and made them into critical realists, there are still, thank goodness, a few things they can hold onto. But before I go into them we have to account for the fact that in the last sentence we see another relapse into a basically naive assumption.

As a matter of course we characterized our subjects, now cured of their naiveté, as "critical realists." But, on closer inspection, isn't that still saying too much? Doesn't this term too still contain an untested presupposition by asserting that we are dealing with some kind of established fact? What is a realist anyway? Can we be absolutely sure that there is an objective world "out there"?

After philosophers had taken naive realism completely apart, in accordance with the rules of their art, they had to acknowledge that far more was buried beneath the wreckage than simple faith in a flawless fit between one's own experience and objective reality. All of a sudden doubts arose whether any such reality existed outside the consciousness of the person having the experience.

When so many qualities previously thought to be inherent in the world itself proved to be purely subjective "psychic" experiences, why couldn't the whole world be nothing more than a fiction of our brains, a mere idea, or a dream, or however else one wished to call such an illusion? Is there actually any way to prove the existence of "extrasubjective reality"?

Philosophers were stunned to realize that it was essentially impossible. Extreme idealism or solipsism (the assumption that only my ego exists and that everything else, including my fellow men, is only a dreamlike projection) is irrefutable, because any one who consistently defends this point of view will naturally interpret any and all proofs of reality—even a brick falling off the roof and landing on him, or a violent attack by another person —as inventions of his own consciousness.

But the solipsistic hypothesis is also unprovable. For logical reasons a proof would be possible only if at least theoretical arguments could be thought up that would contradict the hypothesis, only to be convincingly disproved by it. And hypotheses that can neither be proved nor refuted are not only boring but completely sterile. They produce nothing, no questions, no explanations, no conclusions. That in itself, of course, would not be a sufficient reason for rejecting them. (They could always, despite everything, be "right.") Alongside them, however, there are always other hypotheses that admittedly cannot be proved with final certainty but can at least be considered probable. In that case one is justified in choosing the plausible hypothesis, so long as one maintains a critical awareness that it's only a hypothesis, which in the final analysis can never be proved once and for all.

Among such hypotheses contemporary epistemology would list the assumption that an objective world exists outside our consciousness, that there is any extrasubjective reality, quite apart from the question of whether and to what extent we can know it and quite apart from our own individual existence. As modern cognitive research sees it, all this is—to stress the point once more—nothing but a legitimate assumption.

In principle it cannot be proved. Hence no one can be compelled (on purely logical grounds) to subscribe to it. Still, the decision to do so is a rational one, because among all other alternative assumptions (including idealism) this one is the most likely. As Karl Popper says: "I maintain that realism can be neither proved nor refuted. Like everything outside of logic and basic arithmetic it is unprovable. But whereas empirical scientific theories are falsifiable, realism is not. . . . But one can argue for it, and the arguments are overwhelming."[3]

Thus those realists, standing in the ruins of their naive world view, have strictly speaking turned into "hypothetical realists." They have discovered that their experience in no way correctly "reproduces" the world. They have come to realize that their experience can't even prove the existence of an extramental world, but they have nonetheless determined to believe in the existence of a reality beyond themselves.

This is the situation, it seems to me, that all of us are in. The only exceptions are people who swear by extreme idealism and consequently take us all for imaginary products of their own minds. But they are obviously in the minority. The rest of us don't doubt for a moment that the world has a real existence apart from us, even if we have to admit that we can never prove it.

What it comes down to is that the reality of our everyday world has turned out to be an object of faith. What we had taken to be an unquestionable standard against which to measure everything inside and outside this world that lays claim to reality has been revealed as a pure hypothesis. Recognizing the world as real is not an act of cognition but of trust, a free decision in favor of one possibility among various others, such as idealism.

It is already clear by now that theological claims as to the existence of a "Beyond" cannot be reduced to absurdity simply by pointing to the existence of a concrete, sensuous, experiential world. This objection to religion has no compelling logical force, but for many people who take it seriously it most certainly works on the psychological level.

Nevertheless there is no foundation at all to the demand for objective criteria to establish the existence of the "Beyond" similar to those used for the empirical world. Such a requirement presumes that this-worldly reality is objectively graspable, which is demonstrably false.

Having come this far, what should the hypothetical realists' next question be? They will ask themselves to what extent their senses and intelligence provide them with correct (or "true") information about the external world. They'll want to know how far they have to lower their cognitive ambitions. They have already seen that they can't know the world "just as it is." What is the relationship linking what their senses tell them (and what their mind makes of such data) to the "true nature" of the world around them?

The history of epistemology in the West is one long chain of attempts to find a valid answer to this central question. This is not the place to enter into the details of this history, nor is there any

reason to do so in the context of this book. But one answer that
Kant gave two hundred years ago (in 1781) has some importance
for our line of thought. This answer denied that there was any
prospect of learning anything whatsoever about the "real" (objective) nature of the world outside our consciousness. Kant did
this with a new and extraordinarily significant argument, and at
the same time he reopened an old issue. Only in our time has a
possible solution to this problem emerged, in the form of "evolutionary epistemology," which enables us to see the relationship
between our minds and the "reality of the world" in a completely
new light.

But first things first. Kant answered the question of whether we
have a chance of learning anything about the true nature of
things and the world around us with a flat no. He had discovered
that whenever we perceive or come to know something, our
knowledge does not conform to the objects but the other way
around: objects are evidently determined by our knowledge. In
other words, Kant found that our knowledge (our thought no less
than our imagination) displays innate structures and that what we
experience in the act of knowing is nothing more than the impress of our own mental structures.

It is easy to visualize what Kant was talking about by making
a little thought experiment. One need only consider all the things
in the world that one can "think away." It is not hard, for instance, to imagine that there are no stars. The sun, moon, and
planets can also be thought out of existence, along with the whole
earth (one imagines oneself suspended in empty space). Even
one's own body can be mentally dispensed with, leaving a disembodied consciousness to flit through space. But at this point we
come up against the limits of our conceptual game, because the
self cannot be thought away—otherwise the whole thinking process would come to a halt. And space cannot be thought away
either. It is also impossible to imagine the existence of a bodiless
self floating in empty space without the passage of time; even
disembodied thoughts have to come one after the other. The
upshot of this is that space, time, and the self (or ego) are in all

likelihood necessary preconditions of our ability to have any thoughts at all.

Accordingly time and space are not things that we learn about the world from experience, as all philosophers before Kant had assumed. They are instead structures of thought and intuition, which are part of our thinking *a priori*. Even before we open our eyes for the first time and look about us to see how the world is constituted, it is quite certain that we shall experience it as spatially and temporally structured. Since space and time are innate forms of cognition it is impossible for us to experience anything outside of them. Space and time are, as Kant put it, not the result of experience but the prerequisite for it, judgments that we make *a priori* about the world, inborn prejudices from which there is no escape. Hence we have no right to assume that space and time inherently belong to the world itself, as it is *an sich* (in itself) and objectively, apart from its reflection in our consciousness, which is the only way we can experience it.

We do not experience the world as a space in which temporal processes take place because it is so, but only because our minds transpose whatever imagination or sense perception gives them into spatiotemporal experiences. Human understanding has no other choice. And so the experience of spatiality and temporality teaches us nothing whatever about the world "as it really is."

But this was only the beginning. Kant succeeded in tracking down still more *a priori* forms of cognition. One of the most important of these is causality. Our conviction that every event must have a cause and that sequential occurrences are a chain of causes which bring about certain effects (only to become causes in turn with effects of their own) is not something we first discover by experience. Here too it would only be self-deception to believe that we come upon the principle of causality *a posteriori,* or after the fact, through patient observation of the world, as it were. Kant argues that causality too is an *a priori* form of cognition. It is a prejudice, or prejudgment, that we bring to phenomena, that we stamp them with, so to speak. And so causality too is not a category of the world "in itself."

This world "as it is," apart from the human consciousness that experiences it, remains in Kant's view unreachable. That is more or less the core of his epistemology, which seems essentially compatible with modern "hypothetical realism."[4] At any rate this position appears, in the light of Kant's views, to be reduced to the absolute minimum: Kant himself never denied that there really was an outside world, but he thought that as far as we were concerned it was an unattainable phantom (and so philosophers characterize Kant's approach to reality as "transcendental idealism"). For a convinced Kantian there is no similarity at all between what I think I am experiencing in my perception of the world and what that world objectively is. Instead, the world as mediated to me by my consciousness is, for Kant, a sort of artwork. It is the product of the encounter, made possible by my organs of perception, between the real world and my mind, which discloses a great deal about my innate, *a priori* cognitive forms but tells me practically nothing about reality.

In Kant's view the signals that my senses pick up and from which my brain assembles a picture of the world *do* come from a real external world. But en route to my consciousness they are so radically changed that the final result says nothing more about their source. The order manifested by the world picture in my experience is not a copy of the order in the world itself. It only reproduces the ordered structures of my mental apparatus. Thus far, a summary of Kantian epistemology in modern terms. The argument looks incontestable. It may be unsettling insomuch as it denies any possible access to the reality of the world, but that's apparently something we'll just have to put up with.

But Kant's answer was still not quite definitive. Rather it breathed new and intensified life into an old and rather peculiar problem. If it's true, as Kant maintains, that the subjective image of the world is unrelated in any recognizable way to its objective makeup, then how do you explain the obvious fact that it enables us to live quite satisfactorily in the real world?

Kant saw the problem very clearly (and considered it insoluble). His emphasis on the unrelatedness of the world as it is and the world as experienced leads to a genuine paradox: if causality only exists in my head, why don't I constantly have collisions with reality when I behave in such an un-Kantian fashion as to assume that cause and effect are really connected?

In the two hundred years since Kant this contradiction has become ever more urgent and distressing in view of the continually increasing ability of scientific theories to account for the world as we find it. How was it possible to describe the behavior of gases with mathematical precision, that is, in a way that would allow experimentally testable predictions? What's to be said of our success in observing the atmospheres of distant suns and making accurate calculations with the results obtained? To put it more generally, how does it happen that our minds are undoubtedly able to penetrate at least a little way into the mysteries of nature? And, still more generally, how can we explain the astonishing (and from the Kantian point of view absolutely baffling) fact that our innate structures of thought to all appearances "fit" the structures of the real world? How could they contain any truth about the world if they had not arisen in the confrontation between the individual and the world, but had been issued to us at birth or fallen ready-made from heaven?

But then in a quite unexpected manner science came along and hit upon a discovery that once again relativized this "fit" between cognitive forms and reality. More than a century after Kant, Albert Einstein discovered that the order of the real world does not coincide so perfectly with the order of our mental structures as had been previously believed. *That* is the essential point of Einstein's theory of relativity.

Incidentally, it should be noted that many people when they hear the word "science" still think exclusively of certain kinds of technology—of nuclear power plants, space travel, or laser beams. But those are all examples of applied science, technical products, not science itself. Science is not the same thing as the exploitation by society of scientific insights; it should be understood as basic research, the continuation of philosophy with different means. The repercussions of the theory of relativity on the development of cognitive research provide an especially graphic example of this.

One of the things in the world, as Einstein found out, that does not fit in with the structure of intuitive knowledge is the phenomenon of the constant speed of light. That is, the speed of light (300,000 kilometers a second in a vacuum) is always the same, regardless of the conditions under which the measurement is taken, and hence regardless of the motion on the part of the light source or the observer measuring its speed.

Now that claim is in fact absolutely unbelievable and incompatible with our powers of imagination. Let us assume that a spaceship meets a brilliant meteor shooting through space at 30 or 40 kilometers a second. (Meteors don't actually shine in empty space, but pretend for our purposes that this one does.) It seems perfectly obvious that the meteor's proper motion must add to the speed of the light that streams out in front of it. If the meteor is flying along on a course aimed straight at the spaceship, then anyone who can add and subtract would naturally expect that a physicist on board the ship measuring the speed of light projected by the meteor would not come up with 300,000 kilometers per second but with 30 or 40 kilometers per second *more.*

But Einstein established that this was not so. In the latter case the speed of light would be 300,000 kilometers per second— that's what it would be in *every* case, even if the meteor were racing *away* from the spaceship at the same speed. Or twice or ten times as fast, or even if it reached the speed of light itself (which is only possible in the world of fantasy).

In every conceivable case the measurement is always the same, which is why the speed of light is called a constant. Anyone who objects that he can't imagine this is in the best of company— Einstein couldn't either. Nobody can and nobody ever will. What we have here is a characteristic of the real world that doesn't correspond to the inborn structure of our minds. The idea that the speed of things must obey the usual mathematical laws has been exposed by Einstein as an *a priori* assumption, an innate prejudice.

Because the matter is so important, I want to mention a recent astronomical observation that graphically confirms the constancy of the speed of light as a fact, however unimaginable. In 1974 radio astronomers discovered a binary star with some very striking qualities 16,000 light-years from the earth (a light-year is about 9.4 billion kilometers). After extremely thorough investigation it was shown to be composed of two neutron stars that rotated on their own axes and around each other at an extraordinarily high speed. Neutron stars are extinguished suns that have collapsed into very dense matter. The entire mass of these erstwhile suns had contracted to spheres with a diameter of only 10 to 20 kilometers. One of the two suns proved to be a pulsar; that is, it beamed out radio impulses with the regularity and exactness of a quartz watch.

The precision of radio pulses enabled astronomers to analyze with extreme accuracy not only the proper rotation of the pulsar but also the rotation of both stars about each other (on the basis of the periodic occultation of the pulsar by the twin star). The English physicist Paul Davies now points out in a recent article that this analysis would never have been possible if the speed of light were not just as constant as Einstein claimed.

Davies argues as follows: the two extra-dense dwarf stars revolve around each other at such a blinding rate of speed that it reaches a significant fraction of the speed of light. If this orbital speed were added to the speed of the radio waves that stream out from the pulsar (radio waves are electromagnetic waves just as light is), the effect would utterly frustrate all the investigative skill of the astronomers. In that case the radio pulses emitted on the phase of its orbit during which the pulsar moves toward the earth would travel there at a proportionately higher speed (and the reverse would be true during the phase when it heads away from the earth). Differences in speed of several percent would, however, significantly lengthen or shorten the time it took the radio impulses to reach the earth (some 16,000 light-years away).

Davies calculated that the radio waves emitted on the portion of the pulsar's path toward the earth would arrive several centuries sooner than the waves emitted (during the same orbit) on the way back. Under this circumstance radio astronomers on earth would face the insoluble problem of analyzing a veritable salad of waves, representing countless orbits of the pulsar, whose impulses would be scattered over centuries.

But this is not what happens. Radio telescopes record every single orbit of this unusual binary star system with clarity and precision. Every orbit is registered separately, one after the other, just as it occurred 16,000 light-years away.[5] So here's one more proof that the speed of light (the speed of all electromagnetic waves) really is independent of movement by the light source or the observer, one more proof that the real world has features that we can't imagine because they don't correspond to our innate cognitive forms.

But, fascinating as this discovery is, it does nothing to clarify the problem we've been dealing with in this chapter. Actually it deepens the confusion. Kant showed us that our experience teaches us nothing about the real world. But the very conclusiveness of Kant's arguments made it seem all the more mysterious that despite the evident lack of connection between us and the real world we still manage to find our way with tolerable success. If the notions of time, space, causality, and so forth, don't come from our experience of the world, then how do they enable us to orient ourselves there?

If the individual doesn't acquire these "forms of intuition" in his encounter with the world, then why do they "fit" the world? All that was puzzling enough, but now we have evidence that the fit is not so perfect after all. If it's hard to believe that correct ideas about the world can be inborn, what are we to make of ideas that come from the same source and apparently miss the reality of the world by a hair's breadth?

How to resolve such absurd contradictions? For the last twenty years or so one increasingly clear and convincing approach has emerged which seems to be leading, at last, to a satisfactory answer to all these questions. It dates from the first realization that our innate cognitive structures might not have fallen from heaven quite as directly and abruptly as people had thought, that they might have a long evolutionary history behind them.

Konrad Lorenz was the first one to come upon the idea that these *a priori* structures could well be *a posteriori* structures, that they might possibly delineate experience of the world (thus explaining the mystery of their fit). This solution had looked impossible so long as people kept their eye riveted exclusively on the individual. Konrad Lorenz found the answer: innate cognitive forms really are experiential, only the experience was had not by the individual but by the species to which he or she belongs.[6]

This solution of a basic epistemological problem by introducing the concept of evolution offers fresh evidence that this concept actually does describe a principle which lies at the root of every single phenomenon in the world around us. Furthermore this solution has a special fascination in that it also provides a very interesting explanation for the peculiar fact that our inborn cognitive structures don't quite match the real world with perfect precision but—as Einstein was the first to point out—only approximately.[7] The reason for this faulty agreement between reality and our knowledge is so crucial for understanding ourselves and the place of humanity in the cosmos that we shall have to discuss evolutionary epistemology in greater detail.

14. Einstein and the Amoeba

ABOUT 30 years ago the German behavioral physiologist Erich von Holst discovered that chickens are born with a picture of their species' mortal enemy inside their heads. He proved this with an experiment that gives us a great deal to think about—not, by the way, because it was horribly cruel, but because the opposite was true: the chicken never noticed that it was being manipulated. And that is what must give us pause. What if what was true for the chicken applied, *mutatis mutandis,* to humans?

Von Holst anesthetized some chickens and implanted fine, hairlike wires in their brains. The wires were insulated with an extremely thin coating of lacquer, except for the very end, which remained bare. The implants took without any complications, and the chickens didn't notice a thing (the brain is incapable of feeling pain). The idea behind the procedure was to stimulate their brains in the areas where the wires were implanted. The stimuli were provided by electrical impulses, whose intensity and pattern of progression were commensurate with natural nerve impulses in all particulars.[8]

Under the circumstances the chickens were totally unaware that anyone was doing anything to them, that they were being artificially influenced "from outside." They had been raised by hand and trained to move around freely during the experiment on a little table. They went about, completely relaxed, softly clucking, now and then pecking at little spots on the tabletop as chickens do.

They were calm, that is, until the moment that von Holst or one of his co-workers pressed the button that sent a current (indistinguishable from a natural nerve impulse) through the wire to the bare point deep in chickens' brains. Then the scene on the laboratory table immediately changed. The chickens—and this is the startling thing about this experiment—behaved in a way that was still quite natural for them. But they seemed to feel transported unexpectedly into a situation that had nothing at all to do with their objective environment on the empty tabletop.

Here is a single, quite typical example of the many different results obtained; von Holst has described it as a "behavioral program for defense against an enemy approaching on the ground" and recorded it in a documentary film which is still in existence. The reaction begins a few seconds after the electrical stimulation is administered, with the chicken adopting a characteristic "on-guard" posture. It suddenly stiffens in its tracks, raises itself up, and in a visibly nervous state examines its surroundings, bobbing its head back and forth. A few moments later it appears to have discovered something and stares at a certain spot on the table (which is as empty as ever).

The invisible "something" seems to be drawing closer. The chicken grows increasingly excited and begins to stride up and down the table. It makes fluttering evasive moves in the face of the oncoming "something" that now seems even nearer and hacks away at it mightily with its beak as if spellbound. No doubt about it, the animal feels threatened. It behaves as if danger were approaching on the tabletop and it had to defend itself.

How the scene concludes depends upon the circumstances. The director of the experiment can release the button at any time. In that case the chicken straightens up at once and looks searchingly around. In watching the film one can't avoid the impression that the chicken is perplexed by the sudden disappearance of the danger. Once it has finally satisfied itself that the trouble is gone, it ruffles up its feathers in relief and lets out a triumphant cock-a-doodle-doo—obviously a combative response to the vanishing of the danger.

If the current remains on, however, it can happen that as the inner tension becomes more and more unbearable the chicken will look for a surrogate to vent its feelings on. As a rule this tends to be one of the scientists standing around the table. The films show that the chicken prefers to attack the hands of those careless enough to let them rest on the edge of the table. Evidently the size and position of a human hand propped up there are the nearest equivalent to the appearance of the threatening phantom created in chicken's brain by the electrical current.

But when the switch is still left on and the chicken proves unable to drive away its insidious, illusory enemy, the scene generally ends when the chicken forgets its patiently inculcated manners and flaps off the table with a loud cry. This actually causes the enemy to disappear—by severing the fine wire that created it to begin with.

The experiment can be repeated as often as one wants. Provided the appropriate part of the brain is stimulated, the chicken, will always go through the same program. It is important to realize that the only artificial factor here is the electrical charge simulating a natural nerve impulse. This is merely a triggering mechanism. Everything that takes place afterwards is produced by the animal itself; that is, the entire scene with its complex of many varied behavioral elements reoccurs on the empty table every time the button is pushed.

This clearly demonstrates that we are dealing with an innate, "prefabricated" pattern fixed in the chicken's brain—already present, strictly speaking, in the fertilized egg from which the chicken will hatch. Putting it still more precisely, this behavioral repertoire is a component of the hereditary molecule in the nucleus of the ovum, which contains the complete blueprint of the future chicken. When its brain is mature, this program is contained in the form of a specific pattern of neural connections, which, when activated, causes the sequence described above.

The only novel aspect of von Holst's experiments was the method used to trigger this and many other behavioral programs. The behaviors themselves had long been familiar to ethologists. But that is the whole point of the experiments: they literally lay the evidence on the table that behaviors typical to a given species, such as mating display, grooming, the search for suitable food[9] or the warding off of enemies are rigid, inborn programs, stored up in various specific points of the brain. They are innate experiences.

The biological usefulness of this state of affairs is obvious. Under normal conditions the behavioral program for "defense against an enemy approaching on the ground" is activated by catching sight of such an enemy, a weasel, say, or a marten or a cat. (This is shown not only by observations in the wild but—still more conclusively—by mockup experiments in the laboratory. More on that shortly.) The fact that it is innate guarantees that the chicken will react correctly the first time it meets a weasel. A chicken that had to find out by trial and error what weasels are would obviously have a very slim chance of becoming a parent, and hence its genes would be eliminated from the further evolution of its species. For this reason chickens that can't solve the problem of how to behave toward weasels "right off the bat" simply don't exist.

There is no further need, then, to argue for the extreme practical value of inborn experiences. But at this point an entirely different question arises: where do the chicken's inborn experiences actually come from? How did it get them, if it didn't have them itself? Its problem-solving ability can't just have fallen from the sky.

This question is analogous to the one we addressed vis-à-vis Kant's *a priori* cognitive forms. We are fully justified in saying that evidently the chicken too is equipped with innate forms of knowledge about its world. And these forms are likewise *a priori*, because what it knows about its enemy's appearance and the most effective way of reacting toward it is not the product of earlier experience. The chicken, as we have said, is born with this knowledge, which is literally *ab ovo*. Nonetheless this knowledge fits the real world of chickens. How is that to be explained?

In the case of the chicken the answer seems easy—as it often does after the fact with many basic discoveries that took genius to make. Konrad Lorenz[10] found the answer. True, the chicken has not actually had the experience itself; still, the "ground enemy defense program," like all other innate instincts, is knowledge that arrived in the chicken world *a posteriori*. All of them resulted from concrete experiences of associating with the enemy and all the other recurrent factors that make up the specific environment of chickens. Only these are experiences had not by certain individuals, by this rooster or that hen, but by the species as a whole.

The single but critical obstacle that had to be overcome to find this answer is our silent assumption that the ability to learn from experience, first of all, presupposes consciousness and hence, second, must be reserved for separate individuals. But on closer inspection both these ideas are seen to be pure prejudices. They are suggested to us by the commonplace meaning of the expression "have an experience" and they view the process of gaining information, which is what we are talking about, in an all too narrow anthropocentric framework.

Goethe called the eye "sunlike" because he had intuitively grasped how totally adapted our organ of vision is to the physical qualities of solar radiation. Modern science can confirm in all its details the fact that if this *weren't* the case, we would be incapable of seeing the sun—or anything it shines on. The complicated structure of our eyes has been shown to be the "objective correlative" to all the laws of optics, which science has required centuries to discover. For example, the eye makes allowance for (and practical use of) the dependence of focal distance upon the surface curvature of a refractive medium, the connection between the incident aperture and the depth of field of the projected image, parallactic displacement (resulting from binocular vision) as a gauge of distance, and so forth.

This is to say nothing about the refinements of spectral dispersion of sunlight into the various colors or of the altogether unbelievable increase in photosensitivity to a point that lies far below the "optic threshold," a sort of gauge activated by the retina. A designing engineer with the most up-to-date training in the field would have to give up in despair over tasks of such complexity. And for all its strenuous efforts contemporary science is still a long way from even understanding, much less equaling, the technical solutions achieved here.

But eyes have been in existence for an unimaginably long period of time. The engineering problems involved in the construction of efficient eyes were resolved millions of years before the moment when humans first recognized that the relationship between light and vision *was* a problem and made it an object of scientific curiosity.

But is that reason enough (and here's the crucial question) to deny that this exquisitely adapted organ has transmitted "information" about the outside world to the organism? I don't mean bits of visual information mediated by the eye, but the organ itself as incarnate information (e.g., about the physical characteristics of light, the laws of refraction, the spectral composition of sunlight, and much else). With the acquisition of the eye hasn't the organism also acquired the information contained in the structure of this organ—even if it's incontestable that all this took place without the help of any consciously planning agent?

At this point we should recall once more the evolutionary plasticity that a species can exhibit over the course of time. Evolution means continuous adaptation to ever new qualities and conditions in the real world. Every adaptation, in effect, "reproduces" some feature of the environment. And, finally, this is the reason (to quote Lorenz again) that the hoof copies the flat surface of the steppes, the bird's wing copies the air, and the fish's fin copies the water.

That is what ethologists mean when they say that every adaptation is equivalent to a gain in knowledge of the environment. "Life itself," Lorenz maintains, "is a process of accumulating knowledge."

This is how, to return to von Holst, the chicken got its innate, *a priori* knowledge, which is prior to experience but nonetheless fits the world: the animal recognizes its enemy the first time they meet. Now we understand that what the chicken knows is the result of a learning process in which the species has adapted genetically to the existence of its enemies.

With this explanation it suddenly dawns on us what to make of Kant's innate cognitive structures. We can give an analogous answer to the question of why our *a priori* forms match the real world. Kant gave up hope of ever being able to learn anything about that world, but he knew nothing about evolution and its laws. As soon as we take both of these into consideration, the darkness overcasting this part of the epistemological problem is cleared up. To be sure, we think in terms of causality—innately, *a priori,* necessarily. But this in no way prevents us from learning something about the world, as Kant pessimistically saw himself forced to conclude.

On the contrary. The "causal compulsion" of our minds represents innate knowledge of the world itself, knowledge that was acquired during the evolution of our species by gradual genetic adaptation to natural selection by the environment. And every adaptation reproduces a part of the real world. This is true not merely for hooves, wings, and fins, but also for modes of behavior and structures of knowledge. Thus the category of causality is in truth nothing but an image of the order that actually prevails in the real world. As soon as we take cognizance of the fact of evolution and bring ourselves, the human species, into the evolutionary process, the problem that defeated even Immanuel Kant no longer exists.

This sketch of evolutionary epistemology shows the explanatory power of the hypothesis that our minds are not miraculous gifts fallen from the sky but the provisional result of the same evolutionary history that has brought into existence everything else that fills the cosmos today.[11] Evolution apparently has a central role to play in epistemology, that is to say, at the core of philosophy (not to mention linguistics and some other anthropological sciences). But this point can't be developed any further here, nor can a detailed case be made for this promising new discipline, which combines philosophical and biological arguments.[12]

If our minds bear the impress of structures that are inherent in the world, then we no longer need to rack our brains trying to figure out how mind and world fit together. That question has been answered once and for all. But can we now rest comfortably in the belief that owing to our innate structures of thought and imagination we have reliable knowledge about the world?

At this stage caution is imperative. Before we give full rein to our optimism, one more argument has to be looked into. In the explanatory model derived from von Holst's experiment, an essential element is missing.

As a matter of course we said that the chicken recognized a typical enemy of its species *a priori,* which is basically true. But may we without further ado add the conclusion that the chicken's innate knowledge on this point contains "correct" information about its real environment? That would surely be premature. Before we can reach a judgment, we have to know what it is the chicken actually sees when it "recognizes" the enemy.

In what form does the phantom that fixes the chicken's attention on the tabletop appear? Does the electrical impulse cause the chicken to hallucinate and see a weasel, marten, or cat? And if a marten, say, what sort of a marten, big or little, with light brown or dark fur? If we assume it's a marten, don't we then have to ask how the chicken's innate knowledge can protect it from a cat (or a fox or a polecat) if its phantasm perfectly resembles a marten? Or should we assume that the chicken's brain has stored up in it copies of all the enemies (and their variants) it could possibly meet in the real world during its entire life? That is, martens large, medium size, and small, fat and thin, foxes of every nuance, cats of one color, white cats, tabbies, and so on *ad infinitum.* Clearly there's a problem here.

What *does* the chicken see when it recognizes a "ground enemy"? Surprisingly enough, this question can be answered quite precisely. The method used is that of the previously mentioned mockups or dummies. Ethologists have worked out experiments to determine what cluster of stimuli triggers a specific behavioral program. Their point of departure was the consideration that it would be absurd to presume that all the individual variants of possible enemies could ever be stored up.

It seemed far more likely that there were certain key stimuli, made up of characteristics shared by the entire group of enemies, regardless of the differences between them. But what do martens, weasels, foxes, cats, and other enemies that threaten chickens from ground level (raptors are a different matter) have in common? Three qualities immediately suggest themselves: a furry surface, a pair of eyes that stare ominously at potential prey, and a way of creeping up in a crouched position to attack.

So researchers proceeded to build dummies in which the image of the enemy was reduced to these characteristics. They sewed fur into bundles and fastened glass buttons to the end as eyes. And lo and behold, when the buttons were pointed forwards and the "creature" was slowly pulled on a long wire (to simulate slinking) toward a rooster or a hen sitting on its nest, the birds panicked and went through their whole repertoire of defensive behaviors till they were exhausted.

But that was not all. The critical discovery here was that upon comparing the effectiveness of various dummies (which were modified systematically) it turned out that the chickens definitely "preferred" dummies that combined all three hostile characteristics to any other kind, even those that actually resembled the potential enemy much more closely. For example, whereas the bundle of fur was fully effective in triggering a response, a stuffed weasel, placed at the same time right near the nest, initially startled but after that was practically ignored by the hen, which seems grotesque to a human observer. But the dead weasel lacked a crucial characteristic, namely the creeping movement, and so its realistic appearance did it, so to speak, no good.

And the same pattern holds for other animals, as has been demonstrated in countless instances using the appropriate kinds of dummies.[13] Birds of all species from geese to little songbirds, fish, and insects can all be predictably manipulated once the specific stimuli are transmitted to them by means of the dummies. Their reaction is not only foreseeable but inevitable. The animals are perfectly incapable of resisting such clusters of stimuli.

Researchers have even managed to construct "superdummies," which contain the stimuli responsible for evoking a certain behavioral program to an unnaturally exaggerated degree. In many experiments animals will prefer these dummies to realistic imitations, even when all other conditions are the same.

What then do these findings tell us in connection with our discussion about epistemology? The answer is obvious if we once again ask ourselves, in what sense items of information about the external world mediated to us by inborn experience can be considered reliable. This was the decisive point we had reached in the argument.

We said that Kant's *a priori* categories, although innate, still fit the world because they are the outcome of our psyche's evolutionary adaptation to this world. This is why we felt justified in viewing these mental structures as ·inborn *knowledge* about the external world. After all, for millions of years they helped our biological ancestors make their way successfully in a world that was certainly not without its risks.

But in this context what does "reliable" (or "valid" or "correct") knowledge mean? Since we can't immediately examine this issue with respect to our own situation and the "world in itself," let's turn once more to the animal's situation, in hopes that we may legitimately rate it as analogous to our own. With animals—for instance, with the chicken facing a roll of fur embellished with buttons—the question of reliability is evidently depends upon how you define it. Thus, from the standpoint of the chances for survival or biological efficiency the information that the chicken gets (from the cluster of stimuli presented to it) about the conditions prevalent in its surroundings has to be seen as unsurpassably reliable and valid.

The composition of the stimulus pattern (utilizing the smallest number of characteristics that will apply to all possible enemies) is the only conceivable solution to a mind-boggling problem. What evolution has achieved here is nothing less than an "abstractive generalization," which systematically ignores all details

Examples of experimental dummies. A male stickleback, when shown the two dummies in the upper left, prefers the lower one. At its level of development the crucial feature evoking this response is not lifelike appearance but a specific characteristic—the swollen belly of a female about to spawn. For the same reason a robin will not court a stuffed bird lacking the red patch on its breast but the tuft of feathers shown alongside it, provided it has been colored red. And a fritillary will tirelessly chase a rotating cylinder painted with alternating bright and dark stripes. It does this because at a certain rpm the bright-dark interchange simulates the feature that provokes this reaction in nature: the fluttering about by the female butterfly, which presents in rapid succession views of the bright upper side and dark lower side of her wings.

and individual differences. It should be stressed that this "abstraction" was brought about at a time when the possibility of deliberately making use of the same strategy by psychic means still lay in an unimaginably distant future.

As a biological organism, then, the chicken, with its innate knowledge, is optimally informed about the world. Its information is correct, reliable, helpful, or however one wishes to call it. And since its existence is limited to the biological plane, that settles the matter.

But the situation presents itself to the human observer in a somewhat different light. Relative to the cognitive capacity of an animal such as the chicken we find ourselves on a superior level, a sort of metaplane. From this vantage point the scene, whose every aspect is described for the chicken by the closed system of an inborn behavioral program on the one hand and an objective pattern of stimuli that triggers it on the other, takes on an altogether different appearance.

Jacob von Uexküll, who first applied the term "environment" to animals, analyzed numerous examples of this sort—among others, the extreme case of the tick. This insect needs mammalian blood for its eggs to mature. Hence if the species is to survive, the tick must be able to recognize mammals, to distinguish its sole source of nourishment from all other objects in its environment. For this purpose it has an innate program that causes it to drop down from the branches when its primitive sensory receptors note the odor of butyric acid, which is a component of all perspiration, as well as a temperature rise in the immediate surroundings.[14]

That's all the tick needs. Heat and smell define a mammal with sufficient reliability for the purpose at hand—as proved by the presence of ticks in today's world. The only features of a mammal, whether buffalo or mouse, that get through to the tick are its odor and the rise in temperature its body causes. For the tick these are all it knows on earth and all it needs to know.

If the chicken were capable of making comparisons, it could—and rightly so—consider itself superior to a tick's sort of relationship to the real world. Of course, it would have to admit that the tick's information is both functional (the species is doing fine) and accurate (mammalian sweat and body heat *are* qualities of the real world). But it would probably retort that in comparison with its own world that of the tick was terribly narrow.

The tick's world picture is not false. Even ticks have innate knowledge of the world at their disposal. (Any species that didn't would quickly become extinct.) But their world is only a pitiful fragment. The tick's experience is "true" but, compared with the chicken's, desperately impoverished.

Given our ontological placement several stories higher than the world of chickens, however, we would find it ridiculous if the chicken thought its superiority was definitive. We can clearly see that it has only a relative advantage over the tick. We're spending so much time with this fowl because the chicken can teach us something. Its situation, which we can judge "from outside," not just spatially but ontologically, gives us an answer to our question whether an adaptation that ensures survival is identical to an objectively correct copy of the real world. The answer is unequivocally negative.

We should be careful to avoid speculating about what a chicken "sees" when it experiences the world optically. But it can be demonstrated that it does see, and in a way unimaginably different from our own. Once again it's only because of the anthropomorphic quality of human speech that we use the same word for both cases of optic experience. We take it for granted that all eyes in the world see the same thing, irrespective of the kind of creature in whose head they happen to be.

But we need only think of the previously mentioned experiments to convince ourselves that nothing could be falser than this assumption. A chicken that fears a bundle of fur more than a stuffed weasel obviously doesn't see the same thing we do in what is objectively the same situation. Nor does a romantically inclined robin that turns its attention to a tuft of feathers in lieu of the conspecific belle sitting next to it.

Even animals provided with eyes comparable to our own must have vastly different optic experiences of the world. This was shown in distressing fashion by an experiment conducted some years ago by Wolfgang Schleidt (at the time a co-worker of Konrad Lorenz). He was trying to discover how a broody turkey hen actually recognizes its own chicks. The crucial signal turned out to be the chirping sound made by the young turkeys. This was far more important than any other kind of signal, even an optic one, as Schleidt dramatically proved when, following the typical method for such experiments, he began to manipulate the various triggering stimuli.

The hen, for example, accepted a stuffed weasel as one of its own brood after Schleidt had inserted into its enemy's corpse a miniature loudspeaker that gave out the helpless chirp of a turkey chick fallen out of the nest. But when Schleidt obstructed the hen's ears so effectively that it could hear nothing, it unhesitatingly hacked to death one of her own chicks with its powerful beak, when the chick tried to return to the nest. The hen saw it coming toward her but didn't recognize it. And whatever approaches the nest without identifying itself is fought as an enemy. The hen's innate program leaves her no choice.[15]

Under normal conditions such catastrophes never occur—only when ethologists deliberately confuse an animal's signal pattern, thereby wrecking the order in its environment so as to lay bare the structures of that environment. Normally the inborn limitations of a genetically fixed behavioral program and the stimuli that trigger it provide so high a degree of security that the animal has no chance of breaking out of this framework, where it would be exposed to unforeseeable risks.

The only possible conclusion from all this is that evolutionary strategy distinguishes between an image of the world that is objectively faithful and one that is biologically efficient. Survival and knowledge are two different things. And since you have to survive before you can do anything else, the survival value of an adaptation takes precedence over any of its other qualities, however desirable.

This fact prohibits us from inferring that if an adaptation is functional it must reflect the world objectively, correctly, or completely. Every adaptation is, true enough, a copy of the world. The question is, how accurate a copy.

Up to this point everything has been clear enough. We see the tick, we see the chicken, and in both cases we recognize the hopeless distance separating both creatures from the "truth about the world." The chicken may have a considerable advantage over the tick, but from our standpoint the gap between them shrinks into insignificance. So much so that, despite all the differences between them, we rightly characterize them both as "animals," classifying them apart from the human race. So far, so good. But the level from which we pass judgment on the tick and the chicken (and everything else on earth) is not the ultimate one, not the final authority.

The idea that it is belongs to our vast storehouse of prejudices. The quickest way to realize the insanity of belief in ourselves as the Last Word is to examine the silent assumptions behind it. These are, among others: that after 13 or more billion years the world has reached the end of its development—just now, at the moment which happens to be our own lifetime; that within this cosmic evolutionary process our brains have just now arrived at the highest possible stage of development, which enables us to perceive the world as a whole and pass definitive judgments on it; that while it's obvious that ticks and turkeys and apes can never know anything about cosmology or elementary particles, at the same time the possibility of anything above and beyond our understanding is, of course, excluded; that the world over the horizon of human intelligence is, so to speak, empty; or, in other words, that quite by chance, of all creatures we humans ("the crown of creation") are the incarnate climax of evolution, the point at which the limits of knowledge and the totality of all existing beings meet in harmony.

All one need do is write that down and the senselessness of such assumptions becomes transparent. Fortunately we have more than just circumstantial evidence that this is so. (The history of ideas alone ought to warn us against the illusion that logic alone can free us from our prejudices.) Einstein proved it, empirically and irrefutably. He showed that our innate cognitive forms, our inherited foreknowledge of the world, give us only the roughest image of that world. Even our brain—and this shouldn't surprise us—was not developed by evolution to know the world objectively but to improve our chances of survival.

Thus if we have been getting along well in the world for ages with our inborn three-dimensional representation of space, then this proves that there must be something in the real world that corresponds to this three-dimensional structure. But here too evolution, with its characteristic pragmatism, has settled for a mere approximate solution. This was established by Einstein's unshakable discovery that real space has (at least) one additional dimension.

Hence evolutionary epistemology helps us to understand why our innate cognitive structures match the world. At the same time, however, we see that they show us only a partial, hazy picture of the world. Knowing *this* separates us from all other living creatures on earth. It is the single fringe of reality that we've got our hands on. The world as it is remains, once and for all, unreachable, even for us.

It's only a step, says Karl Popper, from the amoeba to Einstein. He is referring to of the method used by both to solve problems,[17] but the quip can also be taken in a more fundamental sense. The distance between Einstein and the amoeba strikes us, located as we are somewhere between these two extremes, as incalculably great. But if we measure it against the distance still separating all earthly creatures from the reality of the world, it shrinks away to nothing.

Where then has this excursus into ethology and evolution led us? What does it add to the case we've been making? Well, we've at least seen that it was premature to assume that scientific progress had unconditionally condemned any sort of faith. The contrary seems to be true. Though it's correct to say that science labors to gain objective truth, one of the truths it has thus far unearthed is the startling fact that the real world transcends— quantitatively, qualitatively, and in unimaginable ways—the horizon of the knowledge available to us at our current level of development.

Without question, therefore, there is reality beyond the reach of our intelligence. (Humanists please note that in the final analysis this can be proved only with the argument from evolution, i.e., with scientific methods that people continue to decry as "materialistic.") To be sure "beyond the reach" cannot simply be equated with what the Church means by "the Beyond." Still we can now be sure that the existence of the world (presupposing that it's real) does not cancel the possibility of the "other world" that the major religions speak of. That sort of objection cannot be derived from a world whose reality can only be postulated in the form of a decision for a hypothetical possibility.

Bloch's attack on "otherworldly drivel" is justified. This attitude, namely the tendency to shirk moral commitments in the face of the shortcomings of this world by pointing to "eternal life" as the be-all and end-all, deserves the harshest criticism. One-sidedness always misses the truth. But for the same reason the exclusive "this-worldliness," which has long since spread all throughout our society, is not only false, but also, in the most concrete sense, evil.

But before we conclude this second part of the book with an attempt to describe the relationship that might exist between our reality (as a world evolving in cosmic time) and the "Beyond" of religion, we have to look into one more objection. Isn't it contradictory to try to write about something that by definition lies outside our ken? Isn't it better to keep silent about something when there is nothing we can say with certainty? Doesn't science owe its many successes to the way it has consistently stuck to its resolve to disregard everything that can't be weighed or measured, reproduced or at least refuted?

We must now deal with this objection.

15. The Utopia of Positivism

Around the middle of the seventeenth century the Royal Society in London witnessed one of the strangest sessions ever held on the premises of that august body. All the participants were scientists, outstanding scholars, the intellectual cream of the nation. That has to be stressed because what the worthy gentlemen, berobed and bewigged, were actually doing on this occasion might be hard to say at first glance.

They met at midnight and took their seats at a round table. One man arose and, while the others began to murmur Latin incantations, drew a circle with a piece of chalk on the tabletop. Then one of the other men carefully removed a splendid stag beetle from a little box and placed it in the middle of the chalk circle. From that moment on an expectant silence prevailed in the room.

The stag beetle turned a few times to the right and to the left. Then it made its mind up and marched in a straight line out across the chalk line up to the edge of the table, where it stopped. The event filled the learned group with delight. There were bursts of relieved laughter; a few men even cursed. They patted one another on the shoulder, shook hands, and vowed that from then on they would believe only what they could verify with their own eyes.

That scene, or something like it, is what one must envision from the story that the renowned physicist Niels Bohr once told his colleague Werner Heisenberg.[18] Bohr was deliberately vague about whether it was an actual attested fact or simply an anecdote. But that's unimportant, for like every good anecdote this one is true even if it never happened. It sums up the situation that students of nature found themselves in some four hundred years ago when "modern" science was taking its first tentative steps.

True, there had been empirical investigation of nature in antiquity, from Thales and Pythagoras onward. Greek scientists had already satisfied themselves that the earth was a sphere, had made well thought out measurements to determine its circumference, and tried to estimate how far away the sun and moon were. But all these efforts were buried in oblivion as people's interests began to turn elsewhere.

During the Middle Ages everyone's intellectual efforts were directed toward weightier matters. The finest minds strove to explain in what sense Christ was simultaneously true God and true man or how the divine essence was to be understood. The answers to such questions were indisputably far more important than knowing whether the earth's diameter was 12,000 kilometers or 20,000. Salvation didn't depend upon *that,* but it might very well depend upon whether one had rightly understood the conditions to be fulfilled to pass the supreme test of the Last Judgment with some margin of safety. Thus it would be quite wrong—as well as unimaginative—to call this epoch "the Dark Ages" merely because of its spiritual and intellectual priorities.

The medieval period was dark only insofar as its search for truth all too often ended in arrogant, dogmatic hairsplitting, intolerance, fanaticism, and brutal persecution. We moderns, with our characteristic determination to build paradise in the here and now, have already wreaked quite as much havoc as our medieval forebears, and we ought not to judge them too harshly.

Nobody in his right mind would hesitate a minute if given the choice between improving his prospects for eternal life or getting detailed information about the nature of the earth. But that, to oversimplify things a bit, was the decision confronting Western society around the end of antiquity, in the centuries after the death of Christ. Given the same option, we would have done the same.

In the last few hundred years the objectives of intellectual endeavor have changed radically, but this is not because its original goal was necessarily an illusion. The real reason is that it dawned on people that they couldn't get such a rational, scientific hold on God and the other world as they had hoped. This realization is something we can be grateful for.

Anselm of Canterbury and Thomas Aquinas have a place among the giants of Western intellectual history (and not just theology), because their schools of thought, in a labor that took centuries, directly or indirectly opened our eyes to the fact that God is not rationally comprehensible, that he cannot be "proved" in the manner of a logical deduction.

Ever since that crucial point was established theology and science have gone their separate ways. Theologians work at understanding their field in a manner more appropriate to its content. Scientists, now retired (for lack of proper jurisdiction) from the job of pronouncing on Heaven, have at length returned to dealing with the mysteries of the heavens. When they started to resume their proper role, some four hundred years ago, the first thing they discovered was that a Herculean task awaited them. During the long period when they were occupied with "higher things," they had ignored the lowly region of nature, and it had been taken over by a gang of shady characters. Magicians and alchemists, sorcerers, astrologists, and quacks claimed to be in possession of nature's secrets. From the evil eye to horoscopes, from lethal spells to the philosopher's stone, there was no claim too outlandish for some to make and others to believe.

This was a true Augean stable, but how to take the enormous job in hand? The only way was through patient, persevering work with painstaking attention to detail. One after another all the monstrous claims and alleged experiences had to be closely scrutinized. The possibility couldn't be dismissed in advance that one or other of them might be true.

And so the "big issues" were left to the theologians and the philosophers, while the scientists tackled the dirty work. After centuries of bold but fruitless speculation they would now deliberately limit themselves to experiments.

What did a ball actually do when it was rolled down an inclined plane? How did light behave when beamed through glass or water? Why did water boil at a lower temperature on a mountain top than at sea level? To find the answers all the clever disputations in the world were of no avail, nor was paging through the complete works of Aristotle. Only one thing helped: to make the event happen and then to observe it precisely.

And that is what was done to test the "wisdom" of the past. Scientists were no longer willing to take it on faith that there was a secret recipe enabling one to make gold. They got the recipe and tried it out. (The result did little to restore confidence, already shaken as it was, in the world of magicians and faith healers.)

Was it actually the case that vermin could be driven out of the house by reciting a certain formula? That cattle fell sick if you burned a secret sign into the stable door? That crockery would never break if the housewife recited certain words every morning before touching the first plate? The number of such claims and recommended practices was legion. Every one had to be put to the test before the eyes of critical witnesses under conditions that precluded cheating and, as far as possible, error and ambiguity.

Thus the midnight scene in the Royal Society, as reported by Niels Bohr. It may seem farcical to us, but in its historical context it makes sense. At that time it was *not* absurd to discover whether or not a stag beetle's ability to leave a chalk circle could be cancelled by a magic spell spoken at midnight.

The scientists of that day must have had the impression that a kindly spirit had presented them with a passe-partout, a key to unlock all of nature's mysteries: experimentation. At the same time something like a code of honor developed among "natural philosophers," a set of rules that one promised to observe in playing the game of science. After all the disappointments and confusion they were determined to hold on to nothing unless it could be *experimentally* (the new magic word) verified.

The unspoken corollary of this principle was to leave the "big issues"—of God and the other world, immortality and the meaning of life—to other disciplines more competent to handle them.

Scientists had become modest; centuries of vain effort had made an impression on them. Certainty could be had only by renouncing all too ambitious claims and holding on to the concrete findings of the experimental method. Everything else would have to be "bracketed."

Up to this point there are no grounds for complaint, nothing for either a believer or a theologian to take offense at. The idea of criticizing the message of theology was the furthest thing from the minds of sixteenth and seventeenth century scientists. They had explicitly declared themselves incompetent to do so. Furthermore practically all of them, as their writings and remarks to their contemporaries attest, were pious and Godfearing men. The one thing they began to protest against was the attempt to prevent them, with theological arguments or even theologically authorized secular power, from making use of their new experimental method and exchanging the information gained from it.

But further development of science was already in the works, a development that, human nature being what it is, simply had to come. It began when with the help of their method scientists achieved success that exceeded their wildest hopes.

The master key of scientific experimentation opened one door after another leading to the secrets of nature. Take a single example in a single discipline, the deciphering of the "laws" of the solar system. The mere fact that this system, which stretched so far beyond the moon's orbit, followed laws that could be apprehended by the mathematics of sublunary, earthbound creatures, must have struck men and women in that era as a revelation.

We have received this knowledge much as children inherit a fortune they have not worked for. And so we can't really gauge the importance of the discovery that the fixed stars in the sky (previously viewed as an entirely different world, the epitome of everything that transcends the earth) are nothing but an ensemble of suns. Giordano Bruno was the first one to grasp this, thereby hoisting himself into the rank of those intellectual and cultural geniuses on whose shoulders we all stand.

To this Dominican friar we owe the revolutionary perception that the earth when seen from the moon is just as much "up in the sky" as we have always thought of the moon. This was the real turning point, rather than the great discovery by Copernicus, who put the sun instead of the earth at the center of the universe but otherwise left everything as it was. Bruno's radical cosmic relativization, given the ideas prevalent in his day, was a triumph of abstraction comparable in its range and consequences to Einstein's. (Einstein freed us from another variant of the same prejudice, i.e., that reality is identical to the way we experience or imagine it.)

How great Bruno's revolution was can be seen from the fact that not even the so-called revolutionaries of astronomy, whom official historiography still prefers to him, would accept the findings of the brilliant Dominican.[19] Bruno was the first one to draw an unmistakable line of separation between the theologians' heaven and the astronomers' sky. This proved to be a blessing for theology, which was henceforth immunized against the misunderstanding that Jesus' ascent into heaven was a sort of elevator ride. (Bruno's contribution, of course, was not appreciated. He was burned at the stake as a heretic on February 17, 1600.)

In those days observation and experimentation had a truly revolutionary impact. We, the complacent heirs of that legacy, seldom give any thought to such things as the connection between Giordano Bruno's cosmic relativism and a number of seemingly self-evident notions about politics and society, but that connection undeniably exists.

The medieval world picture, with its rigid cosmic hierarchy centered on God, mirrored in the consciousness of the men and women of that age the hierarchical structure of their own feudal society (and at the same time legitimated it as "part of the natural order" or "willed by God"). Similarly there can be no doubt that replacing the cosmic hierarchy centered on the earth (or the solar system) with the world picture of modern astronomy was one of the intellectual conditions of possibility for the idea that all men are equal.

This excursus into the history of science is intended primarily to recall the social and psychological circumstances under which modern science took its first steps. The important point here is that the principled decision not to deal with the "big issues" and to concentrate exclusively on experimental investigation forthwith produced results that far exceeded what anyone had believed possible and, still more, what had been achieved by centuries of bold metaphysical speculation about God and the universe.

And, as we know, astronomy was just the beginning. Physics and biology were not far behind. The secrets of matter and of spontaneous generation, the mysteries of how the world and the human race came into existence, all such things were for millennia only myths that one could at best speculate and wonder about. Now scientists suddenly found themselves in a position to address all these questions quite concretely by means of experiments. But the chief thing was that this new way suddenly yielded results. As scientists proceeded patiently from detail to detail, the age-old questions seemed gradually to be finding answers.

Each individual result was like a tessera in a vast mosaic; it was something of enduring value. Measured against the whole, it may have looked tiny, but it was a "chip of truth." Inevitably the idea arose that it must be possible someday to complete the entire picture. It was only a matter of time and patience till scientists had enough chips in hand to reconstruct the truth.

Intoxicated with their triumphs scientists began to forget the self-imposed limitations they owed them to. The simple rule of bracketing the "big issues" had, to an overwhelming degree, been faithfully adhered to in the beginning and the heavens themselves started to reveal their secrets. Sooner or later the notion was bound to emerge that the other Heaven, a realm whose existence even theologians confessed could never be proved, might be a pure illusion. Might not Heaven be just like all the other unprovable claims that had been so energetically disposed of around the beginning of the scientific era?

Putting it bluntly, was Heaven, was belief in God and the whole phenomenon of religion anything more than a subtle form of the same superstition long since expelled from every other region of human life? Didn't science owe all its indisputable victories to its refusal to accept as true anything that couldn't be objectively, experimentally verified?

Opinions were divided on this question. To this day science is by definition the attempt to see how far man and nature can be explained without recourse to miracles. As a methodological principle this is perfectly legitimate. Any scientist who doesn't strictly observe this rule succumbs to ideological thinking and stands unmasked as a charlatan. The crucial issue here, though, is whether one takes this "positivistic" standard simply as a procedural method or believes in it as an ontological principle. The latter viewpoint became a widespread occupational disease among scientists.

Around the beginning of the nineteenth century Napoleon invited the famous astronomer Laplace to visit him and to explain his new theory about the origin of the solar system.[20] When the learned scientist reached the end of his exposition, the Emperor inquired why God had not come up in the course of it. "Because," Laplace proudly answered, "I had no need of that hypothesis."

Anyone who concludes from this story that Laplace was an atheist doesn't understand science. Maybe he was, but the answer he gave Napoleon proves nothing. No scientist, however pious, could have answered differently. Science is the attempt to explain the world without miracles.

But again, the question is whether one wishes to turn methodology into ontology, to turn the rules of the game of science into the sole foundation of one's vision of the world. Blinded by success, scientists fell victim, for a time at least, to this temptation. The belief spread among them that nothing really existed unless they could lay hold of it with their method. They forgot that limiting science to this method had originally been an act of self-critical humility.

A glaring example of this positivistic attitude is a remark by the

English Nobel prize winner Peter Medawar. Asked by a journalist whether he believed in God, Medawar answered laconically, "Of course not, I am a scientist." The man had been seized by the conviction that whatever he couldn't count, weigh, or describe objectively in some other way did not exist. Here we have method absolutized into faith.

This case further demonstrates that high intelligence is of little value when one is wearing ideological blinders, because when Medawar gave his sneering answer, logical positivism had long been exposed as an unscientific ideology. The credit for refuting it goes to Karl Popper (who for some curious reason is still considered by many people to be a positivist himself[21]).

Positivism is a radical philosophical position that aims at utmost certainty. In its most consistent form, known as logical positivism, it maintains that whatever can't be experimentally tested or logically verified is empty talk. For a logical positivist the ideal "true statement" is represented by the equation $2 \times 2 = 4$.

Such a radical demand leaves reality rather impoverished. The world of the positivist is reduced to a number of tautologies (like the one above or "all pintos are spotted") and a network of abstract logical relationships. Positivists of the strict observance remind one—and not by accident—of the definition of the specialist as someone who knows more and more about less and less until he finally knows everything about nothing. The positivist shows us that if you insist on absolute truth in this world you have to build on a very narrow foundation of truisms.

That may sound exaggerated to some people (though an exaggerated treatment of a subject often clarifies it). But one need only consult the *Tractatus Logico-philosophicus,* which is something like the positivist's bible, to be convinced that its author, Ludwig Wittgenstein, came to an altogether similar conclusion.[22]

The oft quoted line from the *Tractatus,* "Whereof one cannot speak, thereof one must be silent," is the shortened form of the positivist credo. An important but usually ignored fact, by the way, is that this sentence is found not in the body of the text but

THE UTOPIA OF POSITIVISM

in the Foreword. In other words it is a programmatic norm, rather than the result of logical inferences.

This program undoubtedly has an aggressive edge. It's aimed against all the idle verbiage, the ingenious linguistic artistry with which philosophers and theologians all too often used to (and sometimes still do) speak in detail about things that cannot be dealt with so concretely. The *Tractatus* is a reaction (though, of course, there is more to it than that) to such irresponsible use of language, as we can see from a second and equally famous line of Wittgenstein's, "What can be said at all, can be said clearly." This is an unmistakable allusion to the widespread tendency to hint obscurely at momentous meanings that are just not there— something that would become immediately evident if the language employed were "clear."

Thus as a program and as a polemic against certain kinds of linguistic dishonesty the *Tractatus* makes sense—but not as a confession of faith. Taken out of context, apart from the intellectual trend to which it is an understandable and legitimate reaction, it refutes itself by its own unbalanced extremism.

Heisenberg says that the insistence on "dividing the world into what one can say clearly and what one must be silent about" is "nonsensical." Pressed to its logical limit, he argues, this dichotomy would make it impossible for us to understand modern physics, where (for example, in quantum theory) we have long been dependent on images and metaphors.[23]

If we look closely at the *Tractatus*, we may easily find ourselves suspecting that its author was much less of a positivist than many of the people who nowadays invoke his name. Some of Wittgenstein's maxims simply don't fit the familiar stereotypes—such as, "The meaning of the world must lie outside it." (This meaning, presumably, is not something Wittgenstein himself would deny. He simply thought it was impossible to talk about it.) Or, "We feel (!) that even when all possible scientific questions have been answered, our problems with life remain quite untouched."

But while the dyed-in-the-wool positivist would resolutely claim that these "problems with life" have no real existence out-

side our imagination, Wittgenstein came to a quite different conclusion. After reading Tolstoy and the Bible he decided to adopt
an ascetical way of life in service to others, working as an assistant
gardener and village schoolteacher.[24] In contrast to most of his
epigones he obviously wasn't troubled by doubts as to the reality
of that domain, about which we can't make statements with a
probative force comparable to mathematical formulas. He lived
as if he felt obliged to do something for whatever-it-was he believed he had to keep silent about.

Consistent positivism would compel all of us to keep silent
about our problems with life (the "big issues"), just as it would
deprive the physicist of the possibility of talking about quantum
theory. But this is a symptom of its extreme one-sidedness, not
an argument against it. As distasteful as the prospect might be,
it could actually be the case that everything we cannot speak of
in a verifiable way does not, in fact, exist.

But logical positivism is not simply one-sided; it's also self-
contradictory and therefore refutable. As early as 1934, in the
first edition of his major work, *The Logic of Scientific Discovery,* Karl
Popper pointed out that the main presupposition of positivism,
the principle of verification, is untenable. That is because there
are no truths in this world that cannot be doubted. Under critical
scrutiny the principle of verification proves to be mere wishful
thinking. The principle itself cannot be verified, which sends the
whole superstructure of positivism crashing to the ground. Even
the statement $2 \times 2 = 4$ contains no "truth," only a tautology,
the linguistic formulation of an agreement made or accepted by
the speakers themselves.

Of course, this agreement should not be considered arbitrary.
We use the familiar rules of arithmetic so unhesitatingly because
they correspond to the innate structures of our intelligence. But,
as Carl Friedrich Gauss (1777–1855) showed, one can also calculate very well with altogether different rules—just as validly, logically, and convincingly.

The same holds for all other "truths" that we have access to,
not least the truths of science. We can't even apply the stamp of

truth to a law of nature, much less an experimental result or a scientific theory. We could do that only if we were absolutely certain, now and for all time, that no new discovery or new fact could ever arise, leading to a correction, howsoever trifling, of what we now hold to be true. But we never have such certainty, and we never can. Think, for example, of what happened to Newton's theory of gravity. The heavens themselves and all the constellations seemed to conform to it. With its help the eclipses of distant moons could be reckoned in advance just as precisely as the behavior of a stone falling to earth. But for all that the theory of relativity took Newton's calculations, which had apparently been verified "out there" in the cosmos, expanded them, improved them, and put them on a new foundation.

Popper discovered that there are no "true" theories, no "true" knowledge. We can never know what aspect, what details of something that nowadays passes for solidly warranted are going to have be improved and adapted to new findings. For this reason we are in principle incapable of "verifying" even the smallest part of our knowledge.[25] The only thing we can do is try to "falsify" it.

Our sole recourse is to search for facts and arguments that might refute what we think we know, thus enabling us to examine our knowledge, correct it, and if possible better it.[26] We will never possess the definitive truth. As science progresses we come closer and closer (though even this is only a hypothesis) to the truth, but only asymptotically, as it were. We shall never reach the goal.

Thus the positivists lose the Archimedean point, whence they hoped to wrench off its hinges and fling away the existence of anything we can't speak clearly about. The old objection holds good: most things can't be discussed with the same clarity as 2 × 2 = 4, but if we were really obliged to observe silence about those things, we would have to turn into mutes. For, as Heisenberg says, "we can say almost nothing clearly," not even in science. Anyone who finds him- or her-self in a reality that remains hypothetical (Popper calls belief in the world's reality "metaphysical") has no logical basis for dismissing the possibility that a transcendental reality might exist.

The reality we experience in our everyday world is, thanks to long habit, just right for us. That's why we always tend to consider it reality *par excellence,* as the only indubitable reality. But this is a crucial misunderstanding. Our everyday world is supremely excellent merely because we're used to it. It's hard to overcome our faith in its supremacy because of the anthropocentric nature of this bias. This makes it hard to realize that we're dealing once again with a conviction bred into us by evolution.

Innate prejudices are unconquerable, but they can always be recognized as objectively unjustified. No amount of will power can ever make us like the smell of sulfur dioxide. But we have already managed to see that our judgment in this and all cases like it is the consequence of a biologically significant adaptation, and not an objective determination. (A sulfur bacterium would have just as much right to "judge" the same environmental signal in a totally different fashion.)

The archaic element in our central nervous system is so predominant that we are bound to reality chiefly by means of inborn judgments and cognitive forms. But the cerebral cortex, which envelopes these old parts of our brain in the form of a layer of nerve cells only millimeters thick, lifts us so far above all other earthly life that we can see right through our peculiar situation.

Thus epistemology has taught us that we are incapable of knowing the objective reality of the world "in itself" (though we've decided to presume its existence anyhow). And the evolutionary point of view has opened our eyes to the fact that (and the reason why) all mechanisms, structures, and programs, with which we have so successfully made our way in the world thus far, reproduce objective reality (if such exists) only in a way that is both highly simplified and rigorously edited to suit our biological needs.

Given this situation, if we should persist in claiming that the world as we experience it is the only valid and real one, the most comprehensive reality imaginable, then we would be making ourselves as ridiculous, *toutes proportions gardées,* as hens or ticks who made the same claim for *their* experience. The "realist" conjures us to keep our feet on solid ground, but none of us in fact has ever trod it.[27]

We have the Middle Ages to thank for the insight that God and the Beyond can't be proved, that they can't be taken into custody by logic. We have modern science to thank for the knowledge that the possibility of God's existence and of a transcendent reality can in no way be refuted. The positivistic heightening of our desire for truth has led us to the discovery that the reality of our everyday world is no less hypothetical than that of the other world.[28]

Thus, at length, we are once again free to make a decision. How are we to use this freedom? As far as this world goes, we had determined to assume, if only hypothetically, that it was real. But, as Karl Popper showed, an "overwhelming" argument can be made in its favor. Could a comparable case be made in favor of the assumption that the Beyond really exists?

16. The Case for Another World

How LIKELY is it that another reality exists "beyond" our world? No one can stop us from asking this question. Certainly not the realists, so long as they make any pretense to self-critical rationality. Nor can any law of nature, since science has brought us to the realization that the world as we know it cannot be the last word.

The hypothesis that the Beyond exists is at least permissible. But is it probable? The question is a critical one, not just for Christianity, but for Judaism, Islam, and Buddhism, because religion, contrary to the common misconception, is more than just a set of ethical obligations. Moral norms can be justified on other grounds—humanist, Marxist, even evolutionary—and hence are of no use in identifying the specific nature of a "religious" attitude toward the world.[29]

Religion is something else. One can speak meaningfully of a religious position only when it includes the conviction of a transcendent reality. Thus a person is "religious" if he or she takes this reality seriously, is convinced of its presence. (Obviously this is the most general sort of definition of a religious attitude. It bypasses the question of what the person in question may believe in other respects, whether he or she belongs to a specific religion or denomination, and so forth.) Religion in this sense gives rise spontaneously to certain ethical obligations, but these are secondary, the result or expression of religious conviction, rather than its essence.

For this reason all religion needs a transcendent dimension.

Religious language is meaningful only if such a Beyond really exists in some form or other. Now since, as we have seen, the possibility of such a reality cannot be excluded by science or logic, religion can't be denounced as mere superstition.[30] This is saying quite a lot, certainly more than many atheists who consider themselves enlightened and up-to-date would feel bound to concede. But, we may ask, is that all?

This book is an attempt to show that the world picture of modern science not only is no obstacle to a religious interpretation of the world, but actually helps religion by making it possible to reformulate certain statements so as to make them more convincing than they were in their traditional form, with their metaphors and images derived from a long-outdated world picture. The two chapters that follow will try to adduce arguments from modern science that might support the hypothesis of a transcendent dimension.

At first glance this proposal will strike many people as paradoxical, because while science can't preclude the possibility of a transcendental reality, don't its findings undermine rather than bolster the "religious hypothesis"? Hasn't it been settled once and for all, at least since the time of Feuerbach, Marx, and Freud, that the content of religious belief must be considered a series of pyschological projections or, to use Freud's words, "infantile wish fulfillments"?

To this day one hears this thesis defended by many people, including scientists. Isn't religion, they say, simply a "consoling fraud"?[31] A century before Feuerbach, Baron Holbach, a German from the Palatinate, taught in Paris that religion could be explained by its social function: it created an unreal world as a substitute for the human happiness that could not be had in society as it actually was. He added that it was at the same time an extremely effective instrument of domination.[32] The same idea recurs in Karl Marx's familiar definition of religion as "the opium of the people" and "the sigh of the oppressed creature."[33]

Could that be false? What can be objected to in Freud's statement that faith in an omnipotent God, in the final analysis, must be understood as the longing of anxiety-ridden man for the protection of a supreme Father? Hasn't man from the very beginning created God in his own (human) image and likeness? And hasn't he then projected this God into heaven so as to feel a benevolent paternal power spreading its wings over him?

All this is true. Only none of it—something that is, strangely, almost always overlooked—in any way excludes the possibility that religious faith might be directed to something real. There can be no question that God's existence is "devoutly to be wished," nor that this world has per se no intrinsic meaning, so that the existence of a transcendent reality which gave it meaning could only be welcomed, not to mention the desirability of life after death.

But what does this tell us? Obviously it doesn't prove that something must exist just because we'd like it to. And it gives us good reason to be skeptical about our penchant for wishful thinking. But that's all. It absolutely does not mean that, for example, God cannot exist *because* we want him to. This argument has been advanced again and again since the Enlightenment, but, as Hans Küng has clearly shown, it rests on a logical fallacy.[34]

From a psychological point of view, the longevity of this untenable conclusion may perhaps be explained as the expression of an unconscious striving for intellectual integrity, or as a sort of psychic strategy aimed at minimalizing an existential risk, a kind of precautionary measure, inspired by fear of disappointment. Still, we can't help being struck by our willingness to side with the extreme pessimist on this point, with the person who prefers to hope for nothing at all rather than take the slightest chance.

This is striking because in any other area of life such a radical distrust of desirable results would be totally paralyzing. We would never get on the bus to get to work. But the question of how safe our trip to the office or factory will be is evidently less important than the question of God's existence. It looks as if the widespread religious skepticism in our society testifies to its peculiar judgment in matters of existential import.

Be that as it may, we are left with the realization that God cannot be disproved, not even with the help of psychology. Granted, the concretized image of God as a sort of Superfather must be viewed as unconscious wish fulfillment. There's no doubt that the image of a divine kingdom with its graduated hierarchy, from the saints through the angels and archangels, all the way up to the supreme Ruler, is nothing but a "sociomorphic projection,"[35] the naive transferral of familiar social structures into the unknown. And history provides us with a wealth of instances of how effectively religion can be misused as an opiate and an instrument of domination.

But these are superficial aspects of a problem that goes deeper. The heaven theologians tell us about might actually exist, even if naive souls, who take mythological images all too literally, insist on giving it a paradisiacal decor borrowed from the only world they know. The question of whether the word "God" is meaningless—as the Vienna Circle once decreed—or bears some relation to reality has nothing to do with people who happen to connect this word with an Omnipotent All-Father or even with a kindly old white-bearded gentleman observing the earthly scene.

None of these objections explains the phenomenon of religion, which is manifestly a constitutive trait of human nature, in such a definitive, compelling fashion that other explanations must be ruled out of order. At all times and on all continents, in all cultures and phases of history humans have been religious, have either believed in the existence of a reality beyond this world or at least considered it a serious possibility.[36]

However important the conditions of capitalistic production or feudal social structures may have been, they obviously could not have been the source of religious feelings in the Ice Age. On the other hand all epochs of human history have certainly supplied enough occasions for "the sighing of oppressed creatures." But at this point a scientist can't help thinking of an altogether different explanatory model.

Some years ago the well-known American cancer expert Lewis Thomas remarked that he was considerably less intelligent than his own liver.[37] What he meant was that if our liver (and circulatory system, hormonal glands, and many other organs) didn't function "autonomously," if we were forced to direct its operations consciously, none of us would survive for more than a few minutes. Even if we knew about the specific functions of the hundreds and hundreds of enzymes that aid the liver in controlling our metabolism, that wouldn't save us. Because the complex, synchronized interplay of these enzymes with one another and with the constantly changing internal conditions of the body creates problems that infinitely exceed the powers of our conscious brain.

The liver has no brain but it solves these problems every moment of our lives. And as long as it remains healthy, it does this so perfectly that it took a long and tedious investigation to find out what an incredible job the liver continually performs. It learned that job, from the ground up, so to speak, in a training period lasting hundreds of millions of years, in the process called evolution. During that lesson the DNA molecule stored up an increasingly large and ultimately immense abundance of information in a sort of genetic memory bank, which is why the liver can now manage a task that would hopelessly overtax its owner.

Experiences like this lead scientists to listen favorably to the initially odd-sounding idea that there can be "learning" and "knowing" apart from brains. This occurs in the form of information gathering (with a view to handling a specific task) and storing in a mode permitting its recall the instant it is needed. It's hard to describe this molecular process, which occurs with every genetic adaptation, in a language that consistently avoids any terms originally coined with psychic processes in mind.

We had the same trouble in Part I describing the strategies employed by evolution—likewise minus brains and consciousness—to bring into existence living organisms at ever higher levels of organization. There too we talked matter-of-factly about the "creativity" of evolution, about the "free-floating fantasy" of mutations, about the way evolution "tests out" or "invents" things. How can such a vocabulary be justified?

The reply that it's only a matter of metaphors is correct but doesn't get us very far. The question then reads, why do the images and metaphors that so naturally suggest themselves here all come from the realm of psychology? There are, it seems to me, two reasons for this; the first is concrete-factual, the second linguistic, and both are crucial to our case. The first reason is that a genetic connection links the structural peculiarities of evolutionary processes, which are most readily characterized by those metaphors, with the source of the metaphors, because it was evolution that produced our brain and its inborn cognitive forms to begin with.

Hence the patterns of evolution are bound together with individual psychic structures by real similarities, correspondences, and analogies that are anything but accidental. A careful scrutiny of them leads us directly into the labyrinth of the mind–body problem. We shall have to take at least a few steps in this maze, but not until Part III and in a somewhat different context.

But at this juncture we need to look more closely at the linguistic basis for using personal metaphors to describe impersonal processes. The reason, briefly put, is that the sphere of reality encompassed by ordinary language is far, far smaller than we generally think. But this means, among other things, that as soon as we leave the world of everyday appearances behind we either have to invent an artificial language or make do with images from that everyday world, if we wish to talk about the reality encountered beyond its frontiers. "Language was devised for domestic purposes, and we should not expect its usefulness to extend very far beyond the limits of ordinary experience."[38]

We have already discussed the anthropomorphic structure of our language (see the beginning of Chapter 9). "Anthropomorphic" in this context means stamped with the impress of immediate, subjective human experience, which must be either active or passive. This is why we freely speak of the ocean roaring or a tree rustling without reflecting that such phrases are relics of an animistic world picture, which our culture (on the conscious level at any rate) has long since abandoned.

Once again, though, we can see through this inconsistency, but we can't do anything to change it. What we have here, to all appearances, is a genetically determined legacy. Modern linguistic studies suggest that the archaic structure of language is based on an innate program. Simply stated, all children, depending on where they were born, acquire a vocabulary in one or other of the languages spoken on earth. But the syntax of these languages, which stipulates the kind of relations that hold between the things spoken about, is a hereditary given (and accordingly very old) and more or less the same in all human languages.[39]

Some quite pressing social problems arise from the fact that the ongoing development of our genetic constitution hasn't been able to keep up with our cultural development. This discrepancy has got us into a situation that Konrad Lorenz once summarized this way: "In our hand the atomic bomb, and in our heart the same archaic instincts of our prehistoric ancestors."

This disquieting gap exists in language too, where it stretches between the problems posed by contemporary society and the archaic structures of the genetically limited linguistic tools that we must use to solve them. This gap is disquieting above all because language doesn't simply serve to describe, but also in very large measure to determine, the solutions our minds are capable of arriving at. To that extent the structure of our language prejudges the results of our thinking. Hence the fact that this structure rests on an archaic foundation has to be significant.

But back to our main argument. For the reasons given above it is both scientifically permissible and simpler to ascribe imaginative and inventive power to evolution, "without thereby turning it into a rational being."[40] Of course, it would always be possible to start from scratch and describe all over again the objective process one has in mind when, for simplicity's sake, one says that evolution "did" this or that. But this would be necessary only if real confusion would otherwise occur. The anthropomorphic structure of language makes personal modes of expression an incomparably more convenient way of describing things. Only a pedant would take exception to it.

More importantly, we can now understand better why from the

very beginning humans have never been satisfied with any one spoken (and later written) language. The most modern instance of this is the language of scientists. Their "jargon," which most nonscientists can barely comprehend, was not invented to create an air of mystery or to protect the splendid isolation of an elite (though, to be sure, it's occasionally used for that). Rather, the peculiarity of scientific language is due to its artificial character. It has grown and developed to the extent that the various scientific disciplines have pushed beyond the everyday aspect of phenomena and into regions for which ordinary language has no descriptive terms. An extreme example of this is the abstract formulaic language of modern physics. The terminology used by the nuclear physicist has departed most radically from ordinary language, and not by accident. The phenomena and relationships that physics aims to describe are found in a microcosm whose laws have not the slightest resemblance to those in the macrocosm of our immediate experience.[41]

But this is not the first time that humans have generated languages for the purpose of laying hold of areas of reality that ordinary language couldn't reach. The many varieties of art, the diverse artistic descriptions of reality can only be understood as the result of the same need. It's not by chance that we find it so oddly difficult to capture the essence of a painting or sculpture in words, that it's almost impossible, despite the many attempts by intelligent connoisseurs, to say what music *is*. Music too, like poetry or the plastic arts, is an altogether different sort of language that has come into existence along with spoken language precisely because of our need to express those realms of the real world that cannot be grasped with words.[42]

The existence of art, therefore, proves that the horizon of reality stretches further than language's ability to follow. The facets of reality that art describes, comprehends, or makes us conscious of for the first time are thus no less real than those accessible to ordinary language. (And naturally the same holds for the various artificial languages of science.)

We may further conclude that since art is as old as human

culture, extending far back into the dawn of prehistory, people have evidently "known" from the very beginning that language does not suffice to grasp completely the world as they experience it. Actually no one "knew" this back then, not in the sense of reflective knowledge, any more than the individual person, as a rule, knows it today. Once again we're dealing with supra-individual or cultural knowledge.

In a recent and noteworthy book F. A. von Hayek points out that the system of behavioral rules that we call culture originally "in all likelihood contained more 'intelligence' than did man's thought about his environment." Our brain is capable of accepting culture but not of devising it. Hence it is "misleading to consider the individual brain or the individual mind as the keystone in the hierarchy of complex structures produced by evolution."[43]

In the opinion of this renowned political philosopher and Nobel prize winner, the world of culture, like evolution, has a kind of "knowledge without brains." We have already seen why no scientist would find this concept surprising. In von Hayek's view the intelligence of cultural systems has been far superior to the intelligence of individual brains for much of human history. It seems to me that one example of this is the fact that people began to describe their reality by means of art thousands of years before their individually accumulated knowledge enabled them to realize that language alone couldn't handle the job.

But if that's so, are we then entitled to assume that the same is true of mythological language—that artificial language which was developed at the very beginning of history to formulate, among other things, the human insight that this world is not self-explanatory?

The more ingrained a prejudice is, the more time we've had to get used to it, to take it for granted, the harder it will be to correct it. The higher an obstacle, the farther back we have to begin our approach to clear it. That is why I have gone on a relatively long excursus to show why we have the right to respect religion (which, as all historical and transcultural experience teaches us, is a characteristic feature of humanity) and to regard it as a form of supra-individual "knowledge."

The stored up knowledge of past generations can quickly "spoil." What alchemists and astrologers, magical healers and sorcerers once believed has become meaningless to us. And the same holds for the contents of the individual brain. But that doesn't mean that all old knowledge is outdated, even though many people think it has to be that way. They are ready—they may even feel obliged—to throw everything religion says on the rubbish heap of superstition, to reject it as passé. They overlook the fact that the temporal limitations on the validity of knowledge apply only to the concrete contents of individual brains. The latter alone are affected by the progress of knowledge. It's a different story with the knowledge represented by the system of behavioral patterns present in a culture. (Von Hayek's essay is so important because it is apt to spread this idea.[44])

Nowadays, in the spirit of radical rationality we are only too inclined to write off taboos or cultural norms as obsolete and meaningless because we can't see any reasonable justification for them. After that one is easily tempted to brush aside such norms the moment they seem to stand in the way of realizing immediately intelligible goals. We already have some early indications that the consequences of such thinking can be unexpectedly negative.[45]

Under certain circumstances a violation of "mere taboos" or "norms based simply on tradition" can have an unpleasant sequel because, as we noted, that cultural tradition contains a kind of knowledge in many ways superior to individual insight. Not everything that we fail to understand is *ipso facto* groundless. Religion is not just a vestige of our antiquated past.

We have seen that the need for artistic expression conceals a knowledge, which the individual may never become conscious of, that an exclusively verbal description of human reality is inadequate. By the same token we have to regard the religious piety innate in humans as a way of expressing the supra-individual insight that this world is not self-explanatory. Thousands of years before individuals discovered that the world as we experience it is not identical to objective reality, that it cannot, ontologically speaking, be the last word, the human race was already making essentially the same discovery in its cultural history.

As far as the language of myth goes, it helped various periods in their attempt to formulate, record, and hand down this group insight, and so we have to regard it as an artificial language by analogy with the other examples mentioned. Just as words borrowed from ordinary language take on a new role in lyric poetry, so in mythological language words are used in a extra-ordinary manner.

Mythological speech is the attempt to convey with images and allegories something that can't be said without them. It always has its shortcomings, and the state of affairs it describes can be grasped only vaguely ("through a glass, darkly," says Paul), but that's nothing new. Nowadays we have yet another obstacle to face: the figurative, metaphorical connotations of the words, which in mythological statements become the vehicles of the actual information, are dependent upon their particular cultural context. Hence they change over the course of time, and become increasingly foreign to the present-day mentality.

And with the disappearance of the symbolic meanings that convey the message, the literal meanings take on greater prominence. The inevitable outcome of all this is what was originally meant as a mythical picture is misunderstood and read all too literally. That's why our traditional myths (from the resurrection of the dead to the Ascension, from the idea of a faraway heaven to the Christian concept of "Son of God") now threaten, because of the way we handle them, to degenerate into superstition. Anyone who narrowmindedly preserves traditional material in a fixed, irrevocable form (and obliges others to do likewise) is willy-nilly contributing to its destruction.

Realizing this has led some scholars, as we have already seen, to attempt a "demythologization" of the old texts.[46] As a nontheologian, of course, one is bound to be irritated by this term. Taken literally, it doesn't make much sense, because by definition it's impossible to talk about this subject except in mythological language. "Demythologizing," therefore, can only mean the task of transposing mythological formulas, whose symbolism derives from a vanished cultural context and which have thus lost their true function in our society, into images and metaphors capable of "working" in the contemporary context. The collective malaise caused by the gulf between what the churches say and what the world knows is a clear symptom that this task is overdue.

These days we often hear it said that people today are less religious than their ancestors; that they are materialistic, rationalistic, exclusively interested in the world of the here and now; that they never wonder whether there might be a reality transcending it. This is surely not true. If religious feeling really expresses a basic cultural attitude that goes deeper than mere individual knowledge, then it's hard to understand *a priori* why this essential feature of human nature should have vanished, overnight as it were, in our time.

Doubtless it is true that most people in our society live as if they had to find meaning for their existence (if any such meaning can be found) within this world. But are they following their own feelings in this—is it a matter of personal conviction? Or do they rather feel obligated to maintain this stance despite its obvious contradictoriness, because in our cultural setting this is only rational way to behave? It is also true that nowadays there are entire nations whose governments expend every effort (using force if needed) to convert the population to "happiness" as defined by science and rationality. But can we overlook the fact that this process requires coercion?

Similarly, while our society pretends to be rational and enlightened, the concrete behavior of its members speaks a quite different language. Alcoholism and drug addiction, a resurgence of superstitious attitudes and practices, the widespread feeling (especially among young people) of the meaninglessness of life, their seduction by various sects, which can be as absurd as one likes as long as they promise "meaning"—all these are unmistakable withdrawal symptoms. Such phenomena, whose spread throughout the Western world has rightly disturbed so many observers, speak for and not against the assumption that man, even contemporary man, has a very pronounced need for religion.

Of course, the evidence also shows that this fundamental need is not being satisfactorily met. Superstition, a biologist might be tempted to say, is the "vacuum activity" of religion. And when one looks closely at the origin of this vacuum activity, one sees that the churches deserve to be criticized, not the people who join "cults." The task of the churches is to provide legitimate satisfaction for the need to believe. If they were meeting this requirement, all the religion-substitutes would be driven out of circulation.

But the churches serve up the extinct mythological formulas of bygone days—fossils, stones in lieu of bread. And when this fare is rejected, they complain out loud that nobody is hungry.

The partnership between the Church and the world breaks down not because—as, strangely enough, one often hears from both sides—the Church demands too much of people. It's quite simply nonsensical to claim that more and more people nowadays are alienated from the church because they can't bring themselves to submit their desires, carnal and otherwise, to the Church's ethical norms. Just how wide this charge is of the mark can be seen in the well-publicized excesses of the cults. Actually it's altogether unbelievable what privations supposedly modern men and women are ready to endure, what extreme demands they will unhesitatingly comply with (going as far as the gruesome mass suicide in Guyana), if only they can be persuaded that what they're doing has some sort of "higher" meaning.

But obviously the churches are losing their powers of persuasion. The theologians may have the true gospel, but the gurus have the arguments. That's where the problem lies. No less than their ancestors, modern individuals are willing to believe that it's vitally important to lead a life in which religious values play a determining role—with all the practical consequences this entails. But by now it should be clear that contemporary men and women can no longer be swayed by the warning "Obey or the devil will get you."

The only cultural context within which we can try to find symbols to restructure the old, unchanging gospel so as to make it more convincing and accessible to the contemporary mind is that of the modern world picture. And this picture is essentially a scientific one—science being understood not as applied technology but as fundamental research, that broad-based effort to learn about nature which should be regarded as a branch of philosophy.

The symbolic framework into which the content of religion must be transferred if it is to continue to have relevance for us is that of a world evolving through cosmic ages and dimensions. Only the Church itself can do this job, but it holds back out of timidity. Theologians have the (correct) impression that incorporating this modern world picture into their thinking will call into question the familiar mythic formulas entrusted to their care. But in their fearfulness they have drawn the wrong conclusion— since they have clearly begun to identify the traditional mythological form with its contents, they are worried that if they try to replace this envelope, the message itself might evaporate. For this reason I shall once again try to point up those aspects of the modern world picture that not only *permit* discussion of a transcendent Beyond but might even give it a new legitimacy. This attempt must remain inadequate because only a theologian would have the proper training to complete it. But perhaps even such an incomplete effort would at least serve to break the barrier of apprehensiveness that keeps scholars more competent than I in these matters from tackling the job.

Let's talk, then, about the other world.

17. Where Is "the Beyond"?

THIS SORT of discussion is necessarily going to be a rough scramble. There will be countless opportunities to stumble and fall—reason enough, perhaps, to stay at home. But we'll go on the hike anyway, and explain why it makes sense to take the chance.

First, the possible pitfalls. At every step we have to keep an eye on what has been said so far in this book. This means both the arguments that have helped us overcome the prohibition against thinking of the Beyond as a meaningful possibility as well as the limits and conditions we must observe in using our newly won freedom.

The fact that the reality of our world has proved to be hypothetical does render it invalid as an objection against believing in a reality beyond it. But that doesn't leave us free to conceive of any world we choose outside the here and now. The refutation of the principle of verification (the demand that we take seriously only the things whose truth can be positively demonstrated) has liberated us from the positivistic "gag rule." But naturally that doesn't mean we can say just anything about anything.

As a world view positivism has not held up under criticism. As a methodological principle for all scientific research, however, positivistic rules are as indispensable as ever—even though we now realize that in science the exalted concept of "truth" must be replaced by a humbler insistence on "verification." The law still holds that we may not contradict our own stock of knowledge (however uncertain and small it may be), if what we say is to have any meaning.

We have seen, admittedly, that we can use and understand the vocabulary of our language only in the tiny milieu of our everyday experiences (and even there only within limits) and in accordance with the immediate meanings of the words. But even in "naive" usage symbolic connotations play an enormous, though largely unnoticed, role. Still, having determined that the greater part of our speech is metaphorical, we don't have the right to choose our metaphors arbitrarily.

Using evolution in reflecting on our ontological situation creates opportunities, but at the same time it also imposes conditions. Comparing the environments of creatures at varying levels of development reveals how absurd it would be to assume that subjective experience of the world and objective reality coincided for the first time in cosmic history on the lonely heights occupied by . . . us. Great as the distance is separating our picture of the world from the environments of all other animals, it would be ridiculous to argue that reality must end at the frontier of our ability to understand it.

But if the total compass of reality immeasurably exceeds our capacity to imagine it, simply knowing this doesn't enable us to say anything meaningful about the world beyond our ontological horizon. In fact, it would seem at first glance that any such talk is out of the question, precisely because it is "beyond the fringe." Wasn't Kant right when he said that reason "spreads its wings in vain, trying to escape the world of the senses by the sheer force of speculation"?

Nevertheless, I believe there are reasons why contemporary scientists might hesitate to endorse this statement without at least minor reservations. They might find a gap or loophole in the argument. After all, things have changed since Kant's day, when the discovery of evolution and its laws was still a long way off.

The difference between the human condition and the situation of all the other animals, even the highest, is not just that only humans know how far the world they experience lags behind the world "in itself." There is no art or religion among apes, because these, our closest cousins, have no idea that their subjective world is incomplete and wholly relative.[47]

Furthermore men and women can conceive of their world—now recognized as a mere torso of objective reality—as made possible, as underpinned, by a more comprehensive reality that remains forever beyond their reach. Philosophers and theologians have thought this way from time immemorial. That in itself is nothing new. But the discovery of evolution, a continuous process spanning cosmic time and space, has made a crucial difference here.

The order of the world, both of the inanimate universe and, especially, animate nature, has always been advanced as an argument for the existence of a "Prime Mover," a Demiurge or Creator, of God. But for many reasons this "teleological" or "cosmological" argument, which appeals to the ordered beauty of the natural world (or to the mere fact that it exists), is not a satisfactory proof of God.

In the beginning of the world—which scientists too believe in (the "Big Bang" theory)—everything now present in this world must have come into existence, not only time (this was the point of departure for our conclusions at the end of Part I)—but also all of matter and the laws of nature. One of the cornerstones of the Big Bang theory is that before it occurred there was not only no time, but also no space, and nothing else besides. Hence the question so often posed by laypersons, what was there before the Big Bang, cannot be answered. It's misleading to say, as scientists usually do, that the question is "meaningless." It's meaningless only insofar as it's unanswerable. As far as our minds are concerned, before the Big Bang there was *nothing,* in the most radical sense of the word.[48]

But the question is not meaningless, insofar as its unanswerableness doesn't deny anybody's right to assume a cause (necessarily a transcendent cause) behind the beginning of the world. To be sure, we must immediately consider the limited range of the words being used here. In describing this unique case—the origin of the world—"cause" must be understood in a sense fundamentally at variance with the usual one. How can "causality," as we know it, help us to understand an event in which something emerges from "nothing"?

The very category of causality was nonexistent until the world came about. Hence the network of effectual connections (only approximately understood) that we refer to in speaking of "causes" can't have been responsible for the existence of the world. Thus we have to accept the fact that there's no disproving the claim that the world must have had a "ground"—although nothing further can be said about the matter.

So we can't prove that there is a God. And one more thing, given the unique conditions surrounding the origin of this world: we can't prove that its existence isn't purely accidental. Much as our feelings rebel against the idea that the universe is a product of chance, it can't be refuted. But even if one opts for the assumption that the Creator is the cause of the world, the static view of the universe current in Kant's day—and long afterwards—leads to an unpleasant consequence: if the world was made at some vastly remote moment such as we see it today (as creationists and, implicitly at least, vitalists argue), then the Creator's role is limited to a moment as final and unrepeatable as the act of creation itself.

In that case God made the world—and then left it to its own devices. It is, of course, and will always remain his creation. And by the same token the traces of its supernatural origin will make their presence felt in its orderliness, efficiency, and beauty for all time to come. Still this world picture by no means excludes the thought that over the course of time God may have become as remote from his creation as the instant when (for the first and only time) he acted as creator. In other words, this concept of creation inevitably raises the question, how could the creator be concretely, contemporaneously present in such a world?

This question goes a long way toward explaining the unpleasant ongoing altercation between theologians and scientists. If God had to intervene in a finished, self-regulating world every time he wished to be present in it, then every moment that he was "there" would constitute a suspension of the natural laws which under normal circumstances (i.e., in God's absence) let this world function autonomously. This linking of "normal" stretches of time, regulated only by the laws of nature, with moments of divine intervention and at least partial suspension of those laws runs counter to all scientific experience. No theological language, however elegant, abstract, or cryptic, can smooth over the fact that every time science has tested such hypotheses it has found them false.

Nevertheless, theologians who argue in terms of a static world picture have to stick to the interventionist model unless they're prepared to resign themselves to the intolerable notion of an uninvolved, faraway God. This obstinacy forces them to divide the world into two parts, thus curtailing science's jurisdiction. We have already discussed in Part I how this static understanding of creation splits the world into (a) creation, where God still rules and (b) a realm devoid of God.

But the discovery of evolution changed the world, and in such a way that if this dilemma hasn't been totally eliminated, it has at least become more bearable. Here and elsewhere the truth, unsurprisingly, remains beyond our grasp, but the theory of evolution allows us to make some progress along the infinitely long road separating us from that truth.

The world, understood not as a finished product but as ongoing creation, need not be considered closed off in the sense that got us into our dilemma in the first place. It is easier to imagine an immediate relationship to a transcendent reality in a world still evolving than in an already completed one.

I shall not attempt anything like another proof for the existence of God. Despite all the newly available material that might be worked into such a proof, the end result would be no different from earlier efforts. A God who could be "nailed down" with human logic would be just another thing, and that is just what we *don't* have in mind. "A God who 'is' is not," as Bonhoeffer said.

It is easier and less contradictory to talk about the possible influence of a transcendent reality upon our world in the context of evolution. No longer are we dealing simply with a solitary, one-time-only mystery but with one that continues on into the present; no longer simply with an unimaginably remote *creatio ex nihilo* but with the fact that from that moment all the way down to the present and on through the future something is taking place that remains essentially mysterious; with the fact that this world, starting from an elemental, primeval state, has developed into the ordered structures we know today, ultimately bringing forth life, consciousness, and individual intelligence.

Once again it has to be stressed that it is not the specific course taken by evolution which constitutes the mystery. The laws it follows, the molecular-biological and other physical mechanisms it makes use of, are basically accessible to reason. Science has largely, though not of course completely, unraveled them. The mystery is not *how* evolution occurs but *that* it occurs.

Given a world that emerged full-blown from nothingness, the question would simply be: why is there anything at all, rather than nothing? Why does this world and its order exist? That's mysterious enough in itself. But the discovery of evolution has made the marvel even greater, the mystery even more unfathomable. Now we have to ask, in addition: how does it happen that this world is capable of continually producing new forms at ever higher levels of development in an uninterrupted, orderly process?

We know that as far as biological evolution goes the world avails itself of the vagaries of mutations. We have discovered the dialectical interplay of chance and necessity that lies behind every genetic adaptation. We are aware of many other factors directing evolution in accordance with certain laws. But when the question comes up of why do the possibilities that result from all this seem inexhaustible, there is no answer. In this world there is no explanation why evolution didn't long ago come to a halt at one of its completed, self-contained stages.

The fact that evolution exists could be, to repeat, purely accidental. The idea would be repugnant but not logically refutable. The issue is only whether we would be behaving rationally to push our skepticism so far as to accept this hypothesis without taking into account its gigantic improbability.

Thus there is no argument to prevent us from assuming that the order unfolding before our eyes in the evolutionary process embracing both this world and the entire universe reflects an order that exists beyond the frontiers of this world. It is not only permissible but even plausible to work from the presupposition that our reality, which we recognize as such only by freely postulating it and whose order cannot be intelligibly derived from the world itself, is upheld by a more comprehensive system.

We are chained in a cave—for good—and we can see only the shadows of reality on its wall. We never see reality face to face. But are we therefore reduced to "sheer speculation"? Who could challenge our right to reason from the shadows (the only thing we can see) to the reality, without which there would be no shadows?

The concept of evolution, seen this way, gives the mystery a new immediacy. It takes the relationship between our world and the transcendent reality needed to explain it out of the abysmal depths of an unimaginably distant past and brings it back to the present. From the perspective of evolution the connection between these two levels of reality did not take place at one time only, once and for all, the instant the world came into being. A continually evolving world is conceivable only as the result of the living relationship that has existed from the very beginning between the world and transcendence.

This, it seems to me, should induce theologians to pay more attention to the world picture of modern science than, owing to the usual prejudices, has previously been the case. If we have not yet actually trodden theological ground—that territory is reserved for those trained to cover it—our arguments have at least removed some obstacles.

What all this adds up to is that the evolving world is open to transcendence (transcendence is present or living in the world). None of these formulations contradicts modern science, though undoubtedly they do run counter to the image many people still have of science. Nonetheless we now have to get a tighter grip on the concept of transcendence.

Up till now we have used the term indiscriminately for the entire realm that we have to imagine as existing beyond the horizon of the world we experience. It's now time to consider that in a certain sense this definition gives rise to various levels of transcendence, an important distinction for our further reflections.

If we take our subjective cognitive horizon as the basis for defining "transcendent," then in the first instance the word refers to the entire compass of the objective (and not immediately accessible) world: that world "in itself," which spills over the boundaries of our genetically imposed categories of thought. We have already mentioned the reasons supporting the conjecture that this part of the world must be far larger than "healthy common sense" ever dreams.

Imagining the actual presence of this first level of transcendence creates no difficulties. We keep meeting traces of it wherever the artificial sense organs contained in modern scientific instruments run up against the limits of our perceptual field. The same thing happens wherever we succeed, with the help of mathematical symbols, in extending the reach of our thoughts ever so slightly into that domain which nature has forever locked us out of. At such times we learn that the barrier separating us from this larger part of the world may be irremovable but it is not totally opaque.

In this "borderland," then, we encounter compelling hints that the space in which we exist must actually have (at least) one more dimension than we perceive or are capable of imagining. And we have already discovered that this genuinely transcendent four-dimensionality produces distinct, palpable effects in our world—we experience them as, among other things, gravitational forces.

We discover a trace of this level of transcendence in the deepest core of matter in the form of a concrete, empirically detectable paradox: particle-wave dualism. Physicists have had to get used to the idea that on the subatomic plane the term "matter" loses its customary meaning. The elementary particles constituting the atom appear in our world both as particles and as waves, depending upon the method we employ to try to observe them.

These two aspects presumably conceal some kind of unity existing beyond our power to experience it, which we can see only in part, only by indirection, "through a glass, darkly."

Here is where the fascination the star-filled sky exerts on us has its actual roots. The sight brings the insuperable limits of our experience literally right before our eyes. Here we have, right in front of us, the space we cannot imagine as having any bounds but about which we have learned that it cannot be infinite.

Epistemology, evolutionary theory, and modern physics have enabled us to discover that the world in which we live is only a fragment, in all likelihood a tiny one, of the real world, a fragment moreover that mirrors the whole most imperfectly. This discovery, however, is equivalent to recognizing a transcendence of which we knew nothing and which is not quite the same sort of thing that theologians talk about. At issue here is an "inner-worldly transcendence," a phenomenon we had overlooked, ignoring suggestive evidence because of our habitual anthropocentric approach.

The phrase sounds paradoxical, but the contradiction is only relative, because that larger part of the world whose existence we can only infer indirectly is transcendent solely as measured against our own cognitive capacity. The nature of this "Beyond" may seem ever so speculative. Everything we can say about it may remain ever so metaphysical. Still, we have to face the fact that it really exists or, more precisely, that parts of it at least exist today, at this moment, in the form of reality which the senses can perceive and the mind can know.

The notion of immanent transcendence makes sense (and can be confidently distinguished from the religious idea of a transcendence overarching the entire universe) only if we acknowledge that the former too can in principle become part of human consciousness. Once again it is evolution that has opened our eyes to this possibility, for how large a fragment of the world will enter the realm of consciousness depends entirely on what stage of development the conscious subject is at.

More simply put: the history of the development of our brain justifies the prediction that if the human race were given enough time—several hundred thousand years would be the minimum required—more new "centers" would emerge in our cerebral cortex, new regions in the service of functions still unknown and undreamt of.[49] And we can be sure that these cerebral centers would not be reaching into thin air. Rather they would open up parts of the world that up till now have lain "beyond the fringe," in the part of the world that still transcends our experience.

Over the course of evolution, too slowly for us to notice but irresistibly nonetheless, elements of the transcendent world are continually transformed into concrete experience, "face-to-face" reality. (At least this is true on the level of "innerworldly" transcendence.) This is how it was in the past. We cannot doubt that the horizon of our world embraces a larger "slice" of objective reality than did the horizon of the Neanderthals or the australopithecines—not to mention still older ancestors of our race. And this is how it will be in the future—if we get the time.

But over and above that it is reasonable to consider that even today subjective worlds exist at stages of development superior to our own. It would be just one more sign of anthropocentric bias for us to insist that worlds whose cognitive horizon surpasses our own could only be realized if humans are given the chance to outgrow (phylogenetically) the limits of our present-day intelligence.

If we wish to get an objective view of the situation, as close to reality as we can come, we have to put ourselves at a different vantage point. Then we would realize how wrong it is to assume that it all depends on the fate of our species, at what point in the objective world the boundary (which has shifted with the progress of evolution) between subjectively realized reality and "immanent transcendence" finally comes to a halt.

Great as our responsibility is, fortunately it does not as yet extend to the cosmos. We need not be concerned, we alone will not determine whether evolution, whether the history of the entire universe, will be able to run its course to the very end or break off before its time. For this reason it seems wise to presup-

208 OBJECTIVE REALITY AND THE WORLD BEYOND

pose the obvious, namely that if an evolutionary process embracing the entire cosmos had begun an experiment on this one planet, it must have done the same on other planets of the countless billions in the universe.

Once again it is just the old anthropocentric, pre-Copernican prejudice that makes us seriously believe that in this vast, unfathomable universe life and consciousness could have arisen only in one place, our earth. Here for the first time we note a connection that gives us an idea how indispensable the existence of extraterrestrial, nonhuman, intelligent beings really is. We shall have to deal with this point more thoroughly in Part III (along with the scientific arguments in its favor). But first, following our usual method, we have to formulate our case once more.

At this point, therefore, let me summarize what we've said so far in a sort of interim balance sheet. All in all, what can be said about the relationship between our world and the Beyond? Where should we look for that Beyond?

This "otherworldly" dimension surely begins for us a great deal sooner than we have long and uncritically believed. It begins far below the level religions have in mind when they speak of the Beyond. As far as we are concerned it lies deep within the world itself. So long as we indulged in the illusion that we were the "crown of creation," naively imagining ourselves the summit of cosmic evolution, we could go on ascribing to ourselves such a high intellectual rank that where we left off the kingdom of God had to begin.

The truth looks different. In reality we have just reached the lowest rung on the evolutionary ladder, enabling our race to know that the world as we experience it is not the same as "the world" pure and simple. As far as we can see, we are the only life form on our planet to arrive at this insight. Nowhere else on earth do we see any potential for thinking that one's own thought and experience span only a part of reality, that the world in which one struggles for survival is incomplete, a fragment concealing a larger reality without which it would not exist.

On this lowest level the largest part of the world still lies beyond the cognitive horizon. With this we can't help realizing that there must be higher levels of development, that in the course of evolution this horizon grows wider; that is, new, ever broader domains of the objective world are incorporated into subjective reality. Evolution, we said, continually converts transcendence into subjective reality, causes individual knowledge to "grow into" hitherto transcendent areas in an ongoing process that to our time-bound senses seems infinitely slow.

In the long run there is no escaping the idea that there must be many places in the cosmos (too many for us to measure) where evolution for various reasons has made more headway than here on earth, many places where it has by now pressed its efforts beyond generating life, consciousness, and knowledge to new heights, enlarging the realm of subjectivity by annexing regions of which we are still ignorant.

Viewed from outside, meta-physically, all this adds up to a picture of the human situation in which our world is placed, as it were, in the center of a globe composed of many nesting spheres. Each of these spheres is formed by a new ontological level, a stage of cognitive evolution to which all the spheres it encloses are subordinate. The Beyond, the transcendent dimension religion speaks of, could be described as the greatest, most encompassing of all possible spheres, or as the outermost envelope, the highest possible stage of cognitive development, which supports and makes possible all the subordinate realities, since it is identical to objective, definitive reality, to truth itself.

All this is, admittedly, metaphysical speculation, framed in mythological language. But need I say once more that this is the only way to discuss our topic, need I explain why it's legitimate and meaningful to do so?

Still, we can take one part of the picture that I used tentatively to describe what would otherwise be indescribable, and we can fill it out with details from our experience. Thus the picture is not an arbitrary one. What more could we ask for?

We said that evolution, as it takes place within the cosmos, is the mode in which the instant of creation is perceived by our brains. We can now add that this evolutionary process continually transforms transcendence into lived reality in ever new parts of the cosmos in ever more comprehensive ways. Seen from this angle evolution is the movement (not in space but phylogenetic time) of the cosmos as it approaches the Beyond. If so, evolution will have reached its final point when the cosmos coincides with the Beyond. That would be equivalent to the end of the moment of creation.

Thus far the leading points of my thesis, They seem to be satisfactory on a number of counts:

They in no way contradict the world picture drawn by modern science.

They permit us to imagine an immediate, active relationship between our world and the Beyond without violating any laws of nature. We can think of the effect of the Beyond on our world by analogy to the effects of the four-dimensionality of "objective" space, which we can perceive—even though this space lies beyond our cognitive horizon—in the form of gravitational forces.

And they guard against the misunderstandings that we have already criticized in Part I. If we are separated from the Beyond by phylogenetic time (rather than by space), we shall never get any closer to it by turning away from this world but only, whether in life or in death, being part of it. Whether our world will be able to "go the distance" through time up till that last evolutionary step which would bring it face to face with the Beyond—that will depend to no small extent upon what we do or don't do.

None of that runs counter to the world picture of contemporary science, but naturally this fact in itself is not enough. In Part III, therefore, we must address the question whether there are also any scientific findings to bolster our case. In this regard we need to know whether the assumption that cosmic evolution is a finite process makes scientific sense and, if it does, whether there are any clues enabling us to say something about this final point.

NOTES

1. For details see Richard L. Gregory, *Eye and Brain* (New York: 1978).
2. From a television interview (September 2, 1980).
3. Karl R. Popper, *Objective Erkenntnis* (Hamburg: 1973), p. 50.
4. To avoid misunderstanding it should be noted that my outline of the problem of knowledge does not follow strict chronological order. The term "hypothetical realism," for example, was introduced only in the last few decades. My concern in this book, however, is not with intellectual history as such, but with the elaboration of a specific argument.
5. Paul Davies, "Why Pick on Einstein?" *New Scientist* (Aug. 7, 1980).
6. This idea was anticipated by the physicist Ludwig Boltzmann. As early as 1897 he wrote: "We view the brain as the organ or apparatus for manufacturing pictures of the world. Because such pictures have such a practical value in the preservation of the species, the brain developed to a particularly high level of perfection, just as the giraffe's neck and the stork's bill reached an unusual length. . . ." Boltzmann also observed that it was a "logical blunder on Kant's part" to infer from the *a priori* nature of the laws of thought that they were "infallible in all cases." In the same way people had earlier assumed that "our ear and eye were also absolutely perfect because they really have developed to an astonishing degree of perfection. . . . Similarly I should like to dispute the notion that our laws of thought are absolutely perfect. . . . They function no differently from all other inherited habits." (Quotations from Engelbert Broda, *Ludwig Boltzmann* [Vienna: 1955], pp. 106–07.)
7. Einstein's discovery also relativizes Kant's assertion that it is fundamentally impossible to say anything with certainty about extrasubjective reality—although it's still an open question, exactly how the constancy of the speed of light may relate to the nature of the world "in itself." I see this case as one more piece of evidence showing how out of line our educational tradition is in drawing such a sharp distinction between philosophy and basic scientific research.
8. Erich von Holst, "Zur 'Psycho'-Physiologie des Hühnerstammhirns," in *Befinden und Verhalten,* ed. J.D. Achelis and H. von Ditfurth (Stuttgart: 1961).
9. An example: when chickens that have been mechanically incubated and have never seen the food their species typically consumes, namely grains, are given a mixture of spherical and pyramid-shaped grain pellets, they peck at the "natural" (round) bits of feed roughly ten times more often than at the others. (Cited in Gerhard Vollmer, *Evolutionäre Erkenntnistheorie*)
10. Konrad Lorenz, "Die angeborenen Formen möglicher Erfahrung," *Zeitschrift für Tierpsycholgie* 5 (1943); p. 235.
11. See my *Der Geist fiel nicht vom Himmel* (Hamburg: 1976).
12. Vollmer, *Evolutionäre Erkenntnistheorie.*
13. Konrad Lorenz cites a host of examples. See also Irenäus Eibl-Eibesfeldt, *Ethologie: Die Biologie des Verhaltens* (special edition of the *Handbuch der Biologie*), vol. 2 (Frankfurt am Main: 1966); in English translation, *Ethology: The Biology of Behavior* (New York: Holt, Rinehart & Winston, 1975).
14. Eibl-Eibesfeldt, *Ethologie,* p. 344.
15. Wolfgang M. Schleidt *et al.,* "Störungen der Mutter-Kind-Beziehung bei

212 NOTES

Truthühern durch Gehörverlust," *Behaviour* 16 (1960); p. 254.
16. Rupert Riedl adduces many more examples in his *Biologie der Erkenntnis* (Berlin: 1979). A major reason why we have failed so miserably in dealing with the increasingly complex economic and ecological problems of civilization is that the causality of their structures is not linear but "interwoven" and marked by multiple feedback, and hence we can no longer cope with them by means of the analytical strategies we are genetically programmed for.
17. Karl R. Popper, *Objective Erkenntnis*, pp. 273–274. Both Einstein and the amoeba followed the method of trial and error. But while the amoeba was forced to carry out all its experiments in order to find the best hypothesis, Einstein scrutinized his hypotheses *before* applying them. Popper: "I maintain that this consciously critical approach to his own thinking is the only really significant difference between Einstein's method and the amoeba's."
18. Werner Heisenberg, *Der Teil und das Ganze* (Munich: 1972), pp. 281–282.
19. Ten years after Bruno's execution Kepler wrote a characteristic letter to Galileo. (This was after the latter had discovered four moons of Jupiter.) Kepler was very well acquainted with Bruno's writings and had for years been preoccupied with the idea that the stars in the sky, if they really were, as Bruno claimed, suns like our own, might have inhabited planets. But now Galileo's telescope had turned up only Jupiter's moons, and Kepler wrote him with an evident sense of relief: "Had you also discovered planets revolving about a fixed star, for me that would have meant banishment to Bruno's infinite universe." He felt, as he wrote in another passage, "a dark thrill of horror at the mere thought of finding myself lost in the immeasurable universe which that unfortunate Bruno proposed, moved by his irrational enthusiasm for infinity."
   Bruno has never been given his due because of the thoroughgoing suppression of his works by the Inquisition. All his works had to be removed from libraries and in Rome they were publicly burned in front of St. Peter's. After Bruno's violent death his work slipped into oblivion for more than two hundred years. The first official editions and translations did not appear till the end of the nineteenth century. (Sources: H. Brunnhofer, *Giordano Brunos Weltanschauung und Verhängnis* (Leipzig: 1882); Dorothea W. Singer, *Giordano Bruno: His Life and Thought* (New York: 1950).
20. It might be noted that, despite all Napoleon has to answer for, I find it hard to imagine any of today's world leaders asking a great scientist to tell them how contemporary science views the world.
21. The misleading positivitic label was, it seems, applied to Popper by members of the "Frankfurt School," and in the first instance by Theodor W. Adorno. See Adorno *et al.*, *Der Positivismusstreit in der deutschen Soziologie* (Darmstadt: 1969).
22. Ludwig Wittgenstein, *Tractatus Logico-philosophicus* (Frankfurt am Main: 1979); in translation, same title (New York: Humanities Press, 1963). Wittgenstein belonged to the so-called Vienna Circle, a group of scientists, mathematicians, and philosophers who founded "logical positivism" in the early 1920s and from it launched their attacks against "all forms of metaphysics and theology."
23. Heisenberg, *Der Teil und das Ganze*, pp. 279 ff.
24. Quoted in Küng, *Does God Exist?* (Garden City, N.Y.: Doubleday, 1980), p. 94.

NOTES   2 1 3

25. As David Hume had already noted, every conceivable logical step goes from the known to the not-yet-known. But this means that the inductive method, that is, of inferring future experiences from the sum of past experiences—the foundation, in other words, of all scientific theory—cannot be logically validated. To take Popper's oft-quoted example: the fact that for centuries every single swan found anywhere in the Northern Hemisphere has been white does not justify the inductive conclusion that *all* swans are white. The inadmissability of this conclusion was, as it turned out, empirically proved by the unexpected discovery of a species of black swans in Australia. It is likewise impossible to predict with absolute certainty that the sun will "rise" tomorrow morning (it might explode as a supernova, the earth could be destroyed by a giant meteor, etc.). Inductive conclusions, therefore, provide only more or less reliable approximations to the "truth."

26. Popper argues that this "falsification principle" is also the basis of all concrete scientific work. A scientist's most important task, he says, is to subject his own theories to constant critical review. This constant effort at self-correction and refutation brings about scientific progress. Others, notably Thomas M. Kuhn in *The Structure of Scientific Revolutions,* make a rather stronger case that scientific progress does not result from such reforms, as may be seen in the web of defensive hypotheses spun around the "epicycle" theory of planetary motion (designed to protect the Ptolemaic system against heliocentrism). For Kuhn a scientific "paradigm" is not improved through corrections made by its supporters, but replaced by a new and more satisfactory paradigm. Kuhn's point is, historically speaking, well taken. But this does not change the fact that Popper's refutation of the "principle of verification" has laid bare the fundamentally and permanently hypothetical character of all human knowledge, and that Popper's "principle of falsification" (as a logical principle, not as a description of the reality of scientific work) gives an accurate account of the probabilistic nature of scientific theory.

27. Common sense insists on interpreting the contact between my fist and the tabletop as "proof" that both are real. In the same vein one might reply as follows: two clouds that collide with each other and alter each other's shape "prove" that they are mutually impervious. But a bird flies through them both without sensing any resistance. Now my fist (like the rest of my body) and the tabletop belong to the same "world" (regardless of its objective status). Hence in bumping into each other, fist and tabletop merely confirm their belonging to the same level of reality—not how real that level is.

28. Just acknowledging the reality of the world as we experience it constitutes an act of trust. Hence the decision to trust in the existence of God is not so absolutely irrational as some self-styled rationalists think. This is one of the recurrent themes of Küng's *Does God Exist?.*

29. This misunderstanding is apparently widespread among educated people, including scientists. Heisenberg quotes Max Planck: "Science . . . presents us with the task of making correct statements about this objective reality . . . but religion deals with the world of values" (*Der Teil und das Ganze,* p. 116). Moral categories, however, are not limited to religion, and hence unsuitable as a *differentia* in defining religion. Planck by the way offered a terse and unmistakable formula in a lecture in 1927: "Religion is the binding of man to God." (*Religion und Naturwissenschaft,* 3rd ed. [Leipzig: 1938], p. 9.)

30. "Superstition" would apply to any conviction making demonstrably untenable claims.

31. Ernst Topitsch, *Erkenntnis und Illusion* (Hamburg: 1979), p. 228.

32. Quoted in Topitsch, *Erkenntnis und Illusion,* p. 25.

33. Hans Küng (*Does God Exist?* pp. 228 ff.) gives an excellent overview of the development of this argument within Marxist ideology. Marx's proverbial definition of religion as the "opium of the people" imputes no demagogic intentions to the ruling class. For Marx religion is an anodyne used by the proletariat to relieve its inescapable misery under capitalism. Accordingly Marx considers it superfluous to campaign actively against religion: it will die off by itself once communist society becomes a reality. Religion as the "sigh of the oppressed creature" will fade away under socialism because the proletariat will have, so to speak, no more reason to sigh. The aggressive formula that religion is "opium *for* the people" makes its first appearance with Lenin. For him religion is a sort of "spiritual intoxicant," deliberately ladled out by the rulers, "in which the slaves of capital drown their humanity, and blunt their desire for a decent human existence." This attitude naturally creates an obligation to combat the "religious stupefaction of the workers." In view of the continued survival of religion in the Soviet Union and other Communist states one cannot help finding Marx's interpretation more plausible than Lenin's.

34. Küng *Does God Exist?* p. 210. In this passage Küng is referring to the philosopher Eduard von Hartmann.

35. Topitsch, *Erkenntnis und Illusion.*

36. In his book, *Urreligiöses Verhalten und Opferbrauchtum des eiszeitlichen Homo sapiens* (Neumünster: 1974), Alfred Rust establishes that this was true for prehistoric humans, even for the legendary Neanderthal, which takes us back to a time at least 100,000 years ago. Rust distinguishes explicitly here between mere shamanistic rituals and evidence for anticipation of a world beyond.

37. Lewis Thomas, *The Lives of a Cell* (New York: Viking, 1974), p. 66.

38. Gerhard Vollmer ascribes this quotation to Wilkinson without providing any further information (*Evolutionäre Erkenntnistheorie* [Stuttgart: 1975], p. 138).

39. Noam Chomsky, *Language and Mind* (New York: Harcourt, Brace & World, 1968).

40. Immanuel Kant, *Critique of Judgment,* § 68.

41. More than twenty years ago the Frankfurt physicist Hermann Dänzer pointed out that the symbolic character of the language used by the nuclear physicist to describe microcosmic reality displays a "remarkable formal analogy" to the symbolic language theologians use to speak about religion, and that hence religious propositions could not be dismissed as unreal simply on the grounds of their symbolic character ("Das Symboldenken in der Atomphysik und in der Theologie," *Universitas* 22 (April, 1967), p. 367).

42. It is no accident, I believe, that this becomes especially clear with reference to certain examples of modern lyric poetry. These works of art are inaccessible to readings bound by the syntax of ordinary language, which strikes me as the reverse side of the fact that they represent an attempt to address domains of reality beyond the reach of ordinary language.

43. Friedrich August von Hayek, *Die drei Quellen der menschlichen Werte* (Tübingen: 1979).

44. One can only hope that von Hayek's readership will not be greatly diminished by the dubious (in my opinion) conclusions he draws from his thesis.

45. On the evolutionary "knowledge" contained in traditional sexual roles see my *Der Geist fiel nicht vom Himmel*, pp. 211–219. See also Bernhard Hassenstein, *Verhaltensbiologie des Kindes* (Munich: 1973).

46. See Part I, note 58.

47. This point is not contradicted by the studies of the Münster zoologist Bernhard Rensch. Rensch got chimpanzees to finger paint and showed that their work revealed an aesthetic sense. (B. Rensch, "Über ästhetische Faktoren im Erleben höherer Tiere," *Naturwissenschaft und Medizin* 9 [Mannheim, 1965], p. 43.) In this context it's interesting to consider Rensch's thesis that the aesthetic effect of certain specific color patterns and body markings might have arisen from the fact that aesthetic factors are elements of the innate mechanism which triggers mating behavior, and that hence they might play a part in sexual selection. (Darwin once made the same conjecture.) But all this comes from a much earlier stage in evolution than the need for artistic expression mentioned in the text. Presumably the need for an artificial language presupposes the intuitive realization of the inadequacy of spoken language—and hence the existence of such a language.

48. If one accepts the cosmological model of the "pulsating universe," then this statement would hold for the *first* Big Bang.

49. For details see my *Der Geist fiel nicht vom Himmel*, pp. 229–242.

PART III

The Future of Evolution and the End of the World

18. The Ghost in the Machine

THE POSSIBILITIES of evolution are inexhaustible. Having just recognized this as a fact we cannot explain, only accept and marvel at, we now must acknowledge that evolution is basically finite. If cosmic evolution is headed towards a goal, as we hypothesized earlier, then it follows that the evolutionary process must be finite.

We could always take the easy way out. We could point to the finitude of the cosmos and leave it at that. All contemporary treatments of cosmology take it for granted that the world has a specifiable age, even though the margin of error in estimating it is naturally wide. The best guess by the current generation of astronomers puts it at about 13 billion years.[1] But however much distance there may be between the Big Bang and us, the notion that the world must have a certain age contains two critical implications.

There can be no talk of "age" in the face of infinite duration. If time is a line stretching out endlessly back into the past and forward into the future, no one point on it, no one moment can be privileged or truly distinguished from any other. Within infinite time however much change occurs there can never be real history. Instead everything that had been would inevitably repeat itself in a gigantic cycle, and this repetition would go on and on forever. "Must not all of them have existed before? Of all things that can happen, must they not have happened already, have passed and been done? Must we not eternally return?" asks Nietzsche, startled by the prospect of "eternity."[2]

But infinity is not just logically impossible. Everything that scientists have discovered about the world to date tells us it is finite both in time and space. In other words, the world had a beginning, and it will have an end. Belief that the world is infinite now rates as an empirically exploded, prescientific dogma.[3]

Thus we could take the easy way out and simply observe that a universe which is itself finite cannot contain anything infinite. Evolution, therefore, both biological and cosmic, must some day come to an end because it cannot outlast the end of the universe.

But we can't be satisfied with that. The hypothesis presented in this book includes the claim that there is a "ground," a cause inherent in evolution which will bring it to an end. I argue that it will not simply run down, exhausted after playing out (perhaps fruitlessly) a vast but not infinitely vast supply of possibilities. Rather it will reach a point that will cancel the possibility of further development together with any need for it. It will end when a summit of development is reached that will retroactively give all evolution its univocal meaning.

The claim, then, is that evolution has a goal—a statement that scientists might vigorously challenge. One of the most tried and tested axioms of modern biology is that the course of evolution is directed by chance, not by any fixed, predetermined goal.

I have no intention of contradicting this axiom, which is a necessary presupposition of all meaningful research in biology, especially evolutionary biology. The contradiction is only apparent, arising as it does from the various meanings of "goal," which tend to become blurred. If one distinguishes carefully among them, the assumption that evolution has a goal can be harmonized with the conviction that it follows a random, undetermined course.

In using the word "chance" (or "random") we run into the same problem. A closer look at this term has shown us that it would be misleading and one-sided (hence false) to equate it simply with "absurdity" and "chaos." Such a narrow interpretation would prevent us from recognizing that only the factor of chance keeps the universe from turning into a soulless machine spinning around in a rut. Thus chance is not merely a synonym for a lack of order and meaning, but also a prerequisite for freedom and with it the possibility of meaning, paradoxical as that may sound (see Chapter 8).

Both cases are alike. Evolution has no "goal," insofar as it is a genuine historical (i.e., nondetermined) process. Its course is not aimed at a concrete target. There are, as scientists would add, no causes working out of the future.[4] If the history of the earth could be turned back four billion years and everything began anew, starting with the "primeval soup" and the as yet inanimate biopolymers, without a doubt the same results would never be obtained, however many times the experiment were repeated. That would call for the reoccurrence of far too many random events, which together determine the pattern of evolution from moment to moment and thereby give it its historical freedom.

But any time the experiment *was* repeated, one thing would definitely turn up again: life. Everything we know about evolution teaches us that sooner or later it would inevitably give rise to life. The only question is what form such life would take. Even with the aid of hindsight we can discover no compelling reasons why the course of genetic history necessarily had to be the way it was. The developmental sequence from fish to reptile to mammal follows a pattern such that one can never predict in advance what the next step will be.

This means that forms of life might have emerged other than the ones which have so far inhabited the earth; that every fish, reptile, and mammal which became a historical reality thereby barred other potential living forms from appearing. Such possible creatures would surely strike us as strange, bizarre, and perhaps repulsive or frightening as well, because they would be alien to us in a much more radical way than fish, reptiles, or mammals are, with whom we at least share a certain degree of kinship.

But the whole idea is chimerical. We could never meet such beings here on earth because we and they would belong to mutually exclusive ancestral lines. If they existed, we would not. Our current place in nature would be occupied by other creatures who

222 EVOLUTION AND THE END OF THE WORLD

would be related in their own way to everything else living on earth, forming a family tree that would leave no room for other possibilities, including us.

The point of such fanciful thoughts is to break our habit of treating what actually happened in the past as the only conceivable kind of reality, as if there were some natural necessity behind it. We look at the familiar forms, past and present, that evolution has brought forth on earth, and we tacitly consider them the only ones possible.

This bias warps our view of the historical "openness" of all evolution. Its genuinely historical character makes it impossible to predict what forms of life it will cause to develop on earth at some future time (or has already developed elsewhere in the cosmos). There is nothing in the laws of nature to dictate what path evolution must take. Every step it takes is the unforeseeable result of collaboration between the opportunities created by accidental mutations and the selective power of the environment. This is what biologists mean when they speak of the nondirectedness, the indeterminacy, or "aimlessness" of evolution.[5]

All this is perfectly valid. Still, there is a possibility worth discussing here that evolution will simply break off at some point, but not because the universe's final day has come. There is, unquestionably, no goal for evolution to "strive for," no magnet to draw it, no force to prescribe its route or in any other way operate on it from the future. But even so it makes sense to consider the possibility that some distant day evolution might come to its own proper end.

There's no contradiction here. The total openness of the actual path followed by evolution does not exclude the possibility that the process as a whole contains tendencies which, were they realized, would mean a fitting conclusion to it. This has nothing to do with the realization of already present embryonic possibilities, nor with any sort of speculation about entelechy. Evolution is not bound, either by the past or the future, as to what it may go on to produce. Nevertheless tendencies do exist that seem to be peculiarly compatible with the nature of evolution and that it helps to actualize (whatever concrete forms may ultimately emerge) in ever increasing perfection.

One of these tendencies seems to be the generation of life. To understand this we have to recall that evolution is not limited to biology. Biological evolution is only one phase, and a relatively late one at that, of a much more comprehensive process covering the whole cosmos both spatially and temporally. Strictly speaking, we have to say that the whole cosmos *is* evolution, from the first moment of its existence. That's why we need to go back to the Big Bang to see that (and in what sense) the prerequisites were already there for all life, which emerged much later from this event.

The English physicist Paul Davies, mentioned earlier, recently compiled some facts that warrant our making such a statement. The subject at issue is the so-called natural constants, that is, measurable values which we have to accept as a given. We come upon them in nature without being able to specify any reason why they should be precisely this or that quantity. One of these constants is the speed of light, amounting, as we know, to almost exactly 300,000 kilometers per second in a vacuum.

No one knows why light travels at just this rate rather than twice or half as fast. The question has no scientific answer; we can only take cognizance of the fact itself. But we are free to think about the possibility that on some other ontological level superior to our own stage of development there might be a compelling reason why the speed of light should be what it is. Which of the higher cognitive spheres that wrap us round would we have to ascend to in order to find this reason?

We simply don't know where and how our world would change if the reason, and with it the speed of light itself in our universe, were to change. We can only assume that the effects would be serious. There are other natural constants that science has recognized as indispensable to our existence. Again we can only take cognizance of them. We shall never know why they are the way they are. But at the same time we have learned that we ourselves and all other life would disappear from the cosmos, that life could never have arisen in the first place, if at the instant of the Big Bang they had turned out to be ever so slightly different.

Among other examples of this Davies mentions the forces at work within the atom. If, for example, the electromagnetic attraction between the electrons of the atomic shell and the protons in the nucleus was noticeably stronger than it is actually is, then the electron shells would be located correspondingly closer to the nucleus (or even collapse into it). But this would result in atoms with a structure in which the cohesive forces on the surface were altered in such a way as to prevent the formation of stable molecular bonds, including those of biopolymers.

To take another example, if the cohesive forces prevalent in the nucleus were only a tiny bit stronger than they are, then the evolution of the universe would have already come to a halt in a much earlier stage. Conditions would have arisen in the nucleus of the atom that would have speeded up the formation of helium from hydrogen through nuclear fusion to such an extent that the universe would have used up its entire supply of hydrogen during the first phase of its development, and nothing would have been left over for making the stars.

If the heat from the Big Bang had been significantly greater than it was, then the background radiation in the cosmos would be so intense, even today, that the whole sky would give off as much energy as the surface of the sun. There would be no water (in a liquid state) on a single planet anywhere in the universe. It is hard to imagine life arising under such circumstances; in any case there would be no humans.[6]

Thus, summarizing these findings, one could conclude that there is an unmistakable connection between certain constants shaping the structure of the universe and that universe's capacity for producing life. No one looking at the chaotic fireball that marked the world's beginning would have guessed this. And the subsequent stage of the first galaxies and the first generation of stars (consisting almost entirely of hydrogen) gave no hint of such future possibilities. But in retrospect we can't miss the fact that the universe emerged from the Big Bang with characteristics that make it look "tailor-made" for the formation of life.

For some years now cosmologists have readily admitted this connection between the whole and its living parts. They have even coined a name for it, the "anthropic principle."[7] The term "biotic principle" would probably have been a happier choice, since the implications of the word "anthropic" only encourage the old temptation to see humans at the center of all cosmic events.

The cosmos did *not* come into existence "in order to bring forth man." But the anthropic principle shows us that, cold and empty as it was, the universe was our native soil: it constituted the first, crucial condition of possibility for the development of life. Of the infinitely many imaginable structures the cosmos might have had, just one (the only one anywhere?) materialized that made life possible and thus, in retrospect, inevitable.

Such a singular arrangement, such a "convenient accident" affecting the universe as a whole, must arouse a certain suspicion among scientists. How strongly some of them object to it can be noted in the hypotheses that some of them have fallen back upon in an effort to challenge the uniqueness of the case. For example, the American physicist J. A. Wheeler has proposed adopting the working assumption that there are infinitely many worlds (with infinitely many different natural constants), almost all of which, however, are "dead," because the special combination of elements needed for the formation of life never materialized.[8] With the help of this assumption the anthropic principle can be easily trivialized: it is obvious that living creatures who rack their brains over the mysteries of a given universe could only do so in a universe capable of producing intelligent beings.

All we are left with if we accept Wheeler's "infinitely many worlds" is a tautology. I cite this almost desperate sounding speculation, which has no other purpose but to trivialize the evidence that our universe was pregnant with life, only to point out how reluctantly scientists acknowledge such a singular state of affairs. Nonetheless they have been forced to recognize it and to incoporate it into their technical vocabulary under the name of the "anthropic principle."

That expression indicates the connection between the structure of the universe and the emergence of living organisms after billions of years of cosmic evolution, which is what I mean when I speak of the evolutionary "tendency" to produce life. I am trying to get at a link between the beginning of evolution and one of its later results—a link that is no more the unfolding of a seed present from the very start (hence some supposed entelechy) than it is a beeline to a preestablished goal (in the sense of a teleological hypothesis).

The cosmos—or evolution, since from this perspective both are the same—has the tendency to give rise to life. In this universe life is no mere bizarre accident, as Monod argues.[9] Life is typical of the universe. Over the course of cosmic history it *had* to develop.

Another tendency observable in this history is the production of organic structures that bring about an essentially new phenomenon of animate nature within an increasingly broad framework and at increasingly higher levels of development. This is the emergence of complex nervous systems and the accompanying generation of psychic experience, which ultimately grows into consciousness, mirroring both the world and itself.

One can describe the latest chapter of cosmic history, that of biological evolution, as a process that has led to the rise and spread of psychic phenomena in the world of matter. It is beyond dispute that evolution brought mind into the material world. In the final phase (thus far) of development, the previously unknown category of the psychic, made possible by specific material systems, manifested itself alongside matter and energy, which up till then had been the only elements in the world.

There is no doubt that "mind," bound up with the material brain, made its evolutionary appearance on earth in the form of individual consciousness. The question is, How are we to understand this "emergence?" Did evolution "produce" mind? Did it "engender" intelligence with the help of the only material available—atoms, elementary particles, and fields of force—within the limits of natural laws? Or is there any other conceivable connection here?

We need scarcely mention that the question of mind and matter has been one of the basic themes of philosophy ever since philosophy began. It is neither possible nor necessary to survey in detail all the answers proposed down through the ages. As is well known, the last few centuries have brought no major innovations on this score. One standard answer is that of idealism. It maintains that from the very beginning mind has been the only reality, and matter is its product. Materialism takes an exactly opposite tack, claiming that only matter exists, likewise from the very beginning and for all time, and all mental phenomena are the effect of its progressive development. A third group of answers grants equal rights to both categories and concentrates on the problems of the ties between them (theories of "reciprocity" or "parallelism").

The fact that centuries of effort by the greatest thinkers have failed to decide the issue allows us to predict with reasonable security that in the final analysis the question is unanswerable. Taking an evolutionary viewpoint, therefore, we may conjecture that on one of the ontological levels superior to our own something happens whose repercussions are felt in our world as matter and mind—akin to the epistemological situation we confronted in the case of the particle-wave dualism.[10]

Hence we find ourselves once more in a domain where proof and counterproof are of no use. Once again we can only speculate. The mind–matter problem is locked in a philosophical stalemate, which gives us a certain amount of freedom to operate. Still, the impossibility of proving any points here doesn't exempt us from the obligation to support our speculation with convincing arguments.

We can do this thanks to the fact of evolution. A close scrutiny of the way "mind" manifested itself in the course of evolution supplies arguments that, in my opinion, make *dualism* look like the most plausible approach. I shall therefore argue that the category of "mind" (we shall try later on to formulate this concept more precisely) exists autonomously and independent of matter. Putting it negatively, evolution has indeed produced

galaxies, stars, planetary systems, and finally life, but the relatively late phenomenon of the psychic (or consciousness) cannot
be viewed as if it did not exist before the moment of its emergence, as if it were generated by cosmic evolution "out of nothing," so to speak.

A number of arguments, as we shall see, can be adduced in
favor of this position. They are not immediately connected with
each other, but this only strengthens the thesis they will serve to
defend. Before going into these arguments, however, I need to
clear away an obvious objection that might otherwise cause misunderstanding: does not my dualistic notion of mind and matter
contradict the teachings of science? The sciences are dominated
by monistic materialism, and this book claims to present modern
scientific ideas that might help theology formulate its message.

This objection ignores a point made earlier: this book's train
of thought has never paid any heed to the boundaries laid down
by the scientific method, which is unavoidably positivistic. We did
not consider ourselves bound by the positivistic "gag rule."
Hence even before this stage in the argument our inferences and
speculations sometimes wandered beyond the limits of what science can say. We took the findings of science merely as jumping
off points for our reflections, in two ways. First, we acknowledged
them as basic conditions that our statements were not free to
contradict; and second, we used them as guidelines or signposts
to extrapolate beyond the world picture they provide us with.

Here we have to stress something overlooked with surprising
frequency in discussions between monists and dualists: the moment one so much as raises the question of the relationship
between mind and matter, one has already left scientific territory,
because within science's (necessarily and legitimately) positivistic
framework "mind" simply has no status.

Science is knowledge concerning the structure and forms of
the transformation of material systems, as well as the spatial
distribution of various kinds of energy. Within the context of
their work scientists systematically limit themselves to monistic
materialism. This limiting is part of the definition of the disci-

pline to which they have dedicated themselves. Scientific study of
living systems is, as we have said, nothing but the attempt to see
how far one can get in explaining the structure and behavior of
these systems exclusively in terms of their material characteristics.

This is legitimate and, as far as practical research goes, the only
fruitful method. But we shouldn't forget that the issue here is not
an assertion about truth but methodical self-restraint—something many scientists have, unfortunately, forgotten. This leads
to an ideological sort of occupational disease, sometimes resulting in the grotesque conviction that mental phenomena simply
don't exist. Consistent behaviorists explain without batting an
eyelash that their own psychic experience of themselves is a pure
illusion.[11]

Strange as that may sound, it's worth noticing that experimental research based on behavioristic principles has unquestionably
been successful in the extreme. This is important for us because
it's the first indication that the complicated bodily (or neuro-
physiological) processes tied in with psychic experience can be
studied and described without regard to the psychic dimension,
that one can comprehend their biological function without bringing in their psychic aspect. We shall presently discuss what this
means.

Under the circumstances it's a little difficult to understand why
so much weight is attached to scientific arguments in philosophical discussion of the mind–body problem. As a methodological
point of departure (for instance, in exploring some brain function) materialistic monism has an undeniable heuristic value. But
what value are scientific arguments supposed to have in a philosophical thrashing out of the mind–body question? For if scientific method constrains us to work within the limits of material
reality and excludes the mental world by definition, then this
method can't help us with the problem of whether mind exists as
an independent category. When a specific declaration is already
given in the premise, then its appearance in the conclusion
doesn't constitute an argument, but a mere triviality (though, of

course, this still leaves the question open on the philosophical level).

On the other hand, a word has to be said here to defend scientists against the oft-repeated charge of "materialism." This does not refer to the deliberately materialistic nature of scientific method. Rather it imputes to scientists a crude and alien notion of matter, as if they claimed that its movements produced everything that takes place in nature, including mental phenomena. Now some people *have* defended such a "degenerate, mechanistic, brickyard materialism." Ernst Bloch, who coined that phrase, recalls a exemplary case of this: at the 1854 Göttingen Scientific Convention the Swiss physiologist Jacob Moleschott announced that just as urine was an excretory product of the kidneys thoughts were nothing but the excretions of the brain. This prompted the philosopher Hermann Lotze, who was also in attendance, to heckle his colleague, quipping that anyone who heard Moleschott talk could see his point.[12]

Brickyard materialism of this sort, however, was unusual, even in 1854. Today one finds it only outside of science in the work of a few coarse-grained ideologues and as a ghost in the nightmares of certain educated but misinformed people. In the latter guise it is alarmingly widespread. Many of those people would be surprised to learn that in their day Marx and Engels battled vigorously against this primitive variant of materialism.[13]

We have already mentioned that in the hands of the nuclear physicists matter has long since lost all of its supposed crudity. C. F. von Weizsäcker goes so far as to say that "matter, which we can define only as that which complies with the laws of physics, is perhaps mind submitting to objectification."[14] Matter understood this way can no longer be used for the ideological club that many "cultured" individuals still think they can batter scientific "materialism" with. Anybody who throws "materialism" around as a term of abuse is simply betraying his or her ignorance.

Thus there's nothing in principle to be said against the attempt to derive psychic phenomena from matter in von Weizsäcker's sense. Nevertheless I intend to argue here, as promised, for a dualistic approach, building on scientific and, in particular, biological evidence. Let's examine the arguments, one by one.

First of all, an objection must be raised against the basic model employed by biological monists to explain the connection (a genetic one, as they see it) between mind and matter. In their view matter generates from itself the mental phenomena of consciousness and intelligence, along with all other psychic events (feelings, sensory experiences, and thoughts). When asked how that could be, they point to the fact that new and unforeseeable "systemic qualities" have appeared again and again during all previous phases of evolution. Psychic phenomena, the monists conclude, are merely a new category of such qualities, which emerged without any prior warning or transitional stage when the development of matter reached a certain degree of complexity.

What does that mean? Behind the abstract language lies a quite ordinary but nonetheless highly remarkable phenomenon. In countless instances, when previously separate elements come together, they abruptly form new systems. If one mixes hydrogen and oxygen, then instead of two invisible gases one suddenly has a single transparent fluid, namely water. At no time are there any other elements involved except hydrogen and oxygen. And yet the system created when they bond acquires characteristics that neither of them possessed before and that existed nowhere else.

If you join a spool of wire and a magnet in the proper fashion you suddenly confront the phenomenon of electromagnetic waves. When chemical compounds with high molecular weight—the so-called biopolymers—unite in a specific complicated structure, there immediately results the novel phenomenon of life. Life, for the scientist, is a new quality of material systems that have reached the level of complexity necessary for it after a sufficiently long evolution. That's the best answer we have to the question of the link between matter and life.

In general, we get the impression that evolution "progresses" from the fact that over the course of time not only do ever more complex material systems come into existence, but they come up with a continual stream of new features. The only thing monists add to this empirically observable pattern is their contention that psychic phenomena developed the same way, that they are merely a new characteristic of highly evolved material systems. As far as I can see, this is the central argument for biological monism.

But is it conclusive? Doesn't it rather beg the question by assuming what has to be proved? Over the course of evolution matter has undeniably displayed overwhelming creativity. It has produced worlds marked by a beauty and abundance of forms that fill us with wonder. "The stuff that all this has been shaped out of ought not to be despised, as people have usually done," according to Ernst Bloch. We shouldn't underestimate matter—it can be credited with achieving practically everything. Nevertheless I don't think it unfair to deny that matter is responsible for the birth of mind. It strikes me as unacceptable to append psychic phenomena to the prior stages of material evolution as just one more "systemic characteristic."

The monists' argument falls apart because it doesn't hold for a single one of those prior stages. Whatever new quality may have emerged during any one of them, it always remained within the material-spatial dimension. But the phenomenon to be explained here is unique. It cannot be mapped or measured, it cannot be approached except in the mode of subjective experience, it cannot be grasped (and this is a "first") by any empirical method.

In all previous cases the "qualitative leap" (emergence of a new systemic characteristic) occurred within space. But now we are talking about a mysterious phenomenon which, although the most certain of all our experiences, does *not* belong to that dimension. The only thing the psyche has in common with the previous "leaps" is that it constitutes something new.

Another fact to be considered here is that new systemic characteristics appear abruptly, without any "preliminary symptoms." Now you don't see them, now you do. This is so typical of the "qualitative leap" that Konrad Lorenz introduced the term "fulguration" (from the Latin *fulgur*, "lightning") to distinguish it from other forms of innovation.[15]

Life too is the result of an evolutionary fulguration. Of course, with the aid of hindsight one can always point out stages of material development that preceded life and maintain that they were a transition between dead and potentially living matter. But this is not true for the fulguration of life itself. A material thing is either alive or dead—there is nothing in between. (It may be hard, when an organism is dying, to say at any given moment which of the two states is present, but that doesn't alter the point.) This is why, despite all the otherwise immense differences separating them, an amoeba is every bit as alive as an elephant or a person.

But all this is diametrically opposed to the fashion in which mental reality came into the world. One can hardly call it fulguration when it took a good billion years for consciousness to awaken, achieve increasing clarity, and reach our current capacity for self-reflection (not that this may be regarded as evolution's final word). There are no varying degrees of life, but there are, unquestionably, innumerable degrees of mind. We can see this both throughout evolutionary history and at the present moment, in the experiential grids of so many species, each at different levels of development.

These grids are literally worlds apart. The biologist, however, along with the psychologist and the paleontologist, can arrange them in a finely gradated series so that they appear a seamless continuity. No doubt about it, mind did not burst into the world like a flash of lightning, it evolved in a process both steady and tortuously slow.[16]

This feature too raises questions about the correctness of making consciousness just another link in the chain of evolutionary stages that led from primeval hydrogen through cosmic, chemical, and finally biological evolution. In the light of the fundamental distinction just drawn between life and mind, such a notion seems unconvincing, forced, and hastily improvised.

Another objection, different but related, can be urged against the monistic thesis that mind is the product of matter. It ties in with the fact that every step evolution takes is the result of natural selection, which "evaluates" a given number of genetic variants, and that hence each of these steps represents some sort of advantage and to that extent can be considered "opportune."

If consciousness were likewise a product of material evolution, then it would have to be demonstrated that the conscious operation of the so-called higher psychic functions was a practical advantage. Remarkably enough, it turns out to be much harder to prove this than one would imagine.

Hans Sachsse, who originated this argument, introduces it by reminding us that a large (I would say, overwhelming) percentage of our higher bodily functions never enter our consciousness.[17] "Our consciousness has only limited access to our total program." And this is true not merely of simple functions (recall the "intelligence" of the liver) but also for higher brain functions, such as the complex patterns executed by the hands of a concert pianist: through patient exercise these have become unconscious and automatic.

Obviously, once something has been learned it can be shunted to parts of the brain that function unconsciously. In view of this Erwin Schrödinger has attempted to define the conscious portion of brain activity as the one that deals with new experience.[18] An interesting notion, but it doesn't really get us very far: we all know how often the most banal old memories can keep thrusting their obstinate way into our consciousness.

But the core of the argument is that none of the functions we ascribe to the psychic domain would be damaged if it suffered the loss of its psychic aspect, that is, if, like our liver or hormonal system, it operated unconsciously. Sachsse makes this clear in an instructive mental exercise. Imagine, he says, that science has managed to discover the cerebral processes which it rightly believes underlie all psychic experiences. Would we then know anything essentially new about what it means to be conscious?

If we could prepare a list where, "as in a dictionary on one page would be thoughts and on the other the corresponding molecular clusters, we could then mechanically reproduce this connection." We could record it in the form of an algorithm and feed it into a computer, in which all these functions could "run," as they do in our heads. If there were a computer capable of handling such a complicated program, it might perhaps have a kind of consciousness. At any rate many people think it would. The possibility can't be ruled out.[19]

But even if this were not the case, every one of those functions would still proceed smoothly—a crucial point in our argument. The greater our success in pinning down psychic experiences to specific cerebral functions, the more superfluous, apparently, that would make the aspect distinguishing these functions from the unconscious ones—and the more the psychic domain would seem to scientists a mere ghost, an intangible and useless ghost inhabiting the machine of our body, as the English philosopher Gilbert Ryle put it.

In this context I should like to bring up the studies by the American brain researcher Roger Sperry and his colleagues, examining patients whose cerebral hemispheres had been surgically separated. It was revealed that the conscious status of brain functions seems to be linked to the *left* hemisphere (in right-handed people at any rate; with left-handed individuals it's the other way around). Thanks to some ingenious experiments a student of Sperry's named Michael Gazzaniga found that such a patient could perform actions directed by the right side of his brain but wouldn't "experience" them. For example, the patient laughed at a cartoon presented only to his right hemisphere (via the left side of his field of vision). But by his own admission he had no idea why he laughed, because he had not consciously perceived the cartoon that triggered his laughter in the manner of a reflex. This sort of evidence allows us to conclude that the emergence into consciousness (normally effected by the connection to the left hemisphere) even of a reaction such as laughter, which strikes us as so profoundly "psychic," is essentially unnecessary. The laugh can occur without it.[20]

Thus, however successful such experiments may be, we can't get any closer to the precise nature of mind by way of the sciences. That is what Sachsse's argument boils down to, and I agree with him. A discipline premised upon the denial that mind constitutes a peculiar realm unto itself can naturally be of no help in explaining the unique features of mind. I repeat, the exclusion of mental reality from the purview of science is not only legitimate but inevitable. But one has to be consistent and admit that for this reason science can't contribute to the study of consciousness (only of the cerebral functions that accompany it). In the final analysis this is simply a truism.

The "materialist" Ernst Bloch says it best: "The dialectical leap from the atom to the cell, from a physical *quantum* to an organic *quale* by way of amino acids is not so hard to conceive. But from the cell to the idea, from a *quantum,* however organically complete, to a psychic, self-reflecting *quale* is not so easy. If we could walk around inside the brain the way we can inside a mill, it would never occur to us that thoughts were being manufactured there."[21]

In the light of the scientific method consciousness, the whole spectrum of what is immediately given in the psychic experience of the individual is a pure phantom. Scientifically based monism derives from a methodological convention. It does not argue but works from previously established definitions. Whenever it appears, therefore, in the form of an ontological declaration, it is strictly an unproved and unprovable conjecture. It may have some value as a heuristic device for interdisciplinary problems (in experimental brain research, for example), but it can tell us nothing about how mind came into the world.

Now mind—in the shape of individual consciousness—*has* emerged into the world and has evolved toward progressively greater clarity. We've every right to think that it could not have been generated by the evolution of matter. But then we're obliged to show how the connection between mind and matter can be more convincingly described. So now is the time to discuss the arguments for dualism.

19. How Mind Came into the World

WE NEED not invoke Descartes to explain why we are not prepared to view conscious experience of the world and the self as a mere phantom. But we can sympathize with the great French philosopher for believing that the world, whose existence can be only hypothetically inferred and whose image is mirrored only imperfectly in the human brain, does not "reach" us with the same immediate certainty as the psychic experience of the self.

On the other hand one has to agree with the scientists when they point out that the possibility of this experience depends upon the undisturbed functioning of the brain. There is no way of proving that animals too have conscious experience of their environment, their feelings and impulses (and to that extent experience of themselves). But the same is true of the experience of our fellow humans, for everyone except oneself. That is why I presume that animals do in fact have consciousness. Imputing it to them is no bolder than assigning it to other people.

But if consciousness is linked to brain activity, we have to allow for the fact that there are widely different degrees of consciousness. No one can deny that behavioral freedom, the ability to learn and make abstractions, as well as other forms of intelligence, become progressively greater as the brain grows more complex. We may take it for granted that the same thing holds for the degree (the "breadth" or "clarity") of the consciousness that goes along with these functions.

What I mean by more or less clarity or breadth of consciousness is familiar to us from our experience of ourselves. When we
come out of anaesthesia or a deep sleep, in the confusion induced
by alcohol or sleeping pills, we experience conscious states
sharply different, in varying degrees, from our normal waking
condition. We can assume that the possessors of less complex
brains experience comparable or lesser (though at least fundamentally similar) degrees of consciousness.

A more difficult question is, at what point on the evolutionary
ladder should we look for the first glimmer of consciousness?
The difficulty may in part be a matter of definition. Still it's fair
to ask how far down the scale (or how far back in time) we can
go and still find, however "diluted," the phenomenon of creatures' experiencing their own inner states (as feelings of pleasure, aversion, pain, etc.). From looking at the brain of a bee or
a fish we can tell that its owner's intelligence must be less than
that of a mouse, a monkey, or a person. But we have no way of
knowing from the anatomy of their central nervous systems
whether bees or fish have feelings of pleasure or aversion. (Nor
can we tells this by observing the human brain.)

Yet their behavior in life-threatening situations or when badly
injured corresponds closely enough to that of all higher organisms so that we can't rule out the possibility of a sense of self,
however dull, in an elementary feeling of aversion at such moments. On the other hand, all these modes of behavior are not
proof. They are all genetically programmed actions of unmistakable practical usefulness (for flight or avoidance) and they fulfill
their functions even if they are not "experienced" (which would
be quite superfluous as far as their biological purpose is concerned).

Opinions are very much divided on the issue. Even protozoa
are not necessarily excluded from this discussion, though they
don't have so much as a nervous system (they can process stimuli,
however). Decades ago the zoologist Oswald Kroh noted that the
amoeba displayed "determinacy of behavior on the basis of individual condition," an impression of spontaneity, and the ability

to orient itself to the most varied environmental stimuli. This led Kroh to attribute "genuine mental accomplishments" to the amoeba, while still labeling it "a long way from any sort of consciousness"—which of course takes back most of what one imagines under the heading of mental accomplishments. Some scientists would go even farther back, granting a "preliminary stage" of consciousness (in the form of "protopsychic" qualities) to matter, even to elementary particles. At any rate a zoologist as eminent as Bernhard Rensch is willing to defend this position.[22]

The question of the psychic frontier is an interesting one, but it can be left unanswered here. One point beyond dispute is that as matter evolved mind went through a process of increasingly more elaborate development. The history of evolution is the same as the history of the emergence and spread of mind in the evolving material world.

Before we can measure the full implications of this fact, we must address the question of how to envisage the relationship between material evolution and the appearance of mind in the guise of individual consciousness. We have seen why the monistic claim that evolution generated this phenomenon along with everything else is not satisfactory. How then are we to understand the statement that the course of evolution equals the increasingly extensive manifestation of mind in the material world?

We can facilitate discussion of this difficult question if we treat it as concretely as possible. The most concrete case of the link between mind and matter that we know of is the one between the brain and its "psychic" functions, so called because they operate consciously. How to imagine the connection between our brain and our thoughts? If the brain doesn't produce them, where do they come from? And why can't we think them without our brain?

Even put this way the question remains hard enough. Even before we start we can say that it can't be given a final answer. It deals with a relationship whose ultimate basis has to be (at least) one ontological level above our present cognitive capacity. Nevertheless we can speak meaningfully about the problem. Nothing we can say, though, could ever be considered proof. Just as the position taken by the monistic materialists can't be conclusively refuted, so we can't argue conclusively for dualism. But at the same time, just as there was convincing evidence that kept us from adopting monism, so evolution provides us with evidence that supports the dualistic approach and shows how the brain–thought connection can be given a plausible metaphysical explanation.

Time and time again in this book we have used terms one generally thinks of as psychological in unusual contexts. We described the mechanism for hereditary transmission as the species' "memory." We spoke of the "free-ranging fantasy" of the mutational principle. Genetic adaptation struck us as a feat of intelligence, and the fit between an inborn behavioral program and the cluster of signals from an experimental dummy that triggered it as the result of an abstraction.

We recall Konrad Lorenz's remark that life itself is a process of acquiring knowledge; that a species' genetic adaptation copies the environmental conditions to which it is adapting; and that even before reaching the biological portion of its history evolution tests, selects, evaluates, tries out, and looks for solutions.

Of course, we shouldn't jump to conclusions. It's always possible that our choice of words will prove to have been dictated by linguistic necessity. It could be that the only way to describe processes occurring *outside* the ordinary world is with language that comes from *within* it. Gerhard Vollmer rightly points out that it would be highly questionable to take the immunobiologists literally when they speak of the reaction between antigen and antibodies as molecular "recognition."[23]

But this use of the pathetic fallacy is by no means accidental, and we can guess why. All the examples just cited refer to fundamental evolutionary mechanisms, to functions therefore that shaped the course of evolution and ultimately produced the human brain. So is it any wonder that the product bears the impress of the structures that formed it? The way we look at the world is more or less predetermined.

Hence the similarity between the process of information acquisition and storage by the genotype of any species on the one hand and individual learning on the other is not a matter of chance. The similarity is based on kinship, on a concrete genetic link: the same strategic discovery made by evolution long ago on the molecular level (which spared it the trouble of always having to redo the same job all over again from the beginning) was put to use eons later by being placed at the disposal of the individual.

This is no mere abstraction and has been used as a heuristic principle in practical research. It's true that the hopes of explaining the molecular biological mechanism at the root of individual memory, which burned so bright in recent decades, have lately for various reasons been dimmed. But we can regard it as certain that such a mechanism does exist, buried somewhere in our brain. And, thanks to the evolutionary connections mentioned above, we can confidently predict that if found it will prove to be closely related, to say the least, to the molecular process of storing hereditary information.

When Konrad Lorenz describes life as a process of acquiring information, he's not simply playing with words. When we talk about selecting, evaluating, testing, learning, remembering, and so forth, we are referring to real mechanisms and strategies that developed long before individual consciousness was even a possibility. In Lorenz's own words, ". . . the phylogenetic process, leading to the formation of meaningful, species-preserving structures, is analogous to individual learning in so many areas that we needn't be surprised if the end results of both are often bewilderingly similar. The genome (system of chromosomes) contains an incredibly rich supply of 'information' . . . , which is accumulated through a procedure closely akin to learning by trial and error."[24]

If we recall the genetic chronology of the connection between the "psychic" achievements of evolution and the conscious activities (which we describe with the same terms) taking place in our heads, then suddenly it dawns on us: we've been taken in once again by the anthropocentric prejudice that would have us believe that every causal chain begins with us.

The strategies of testing by trial and error, of selecting by means of a criterion, the storing up of items of information that have been "screened" and "passed" by those strategies, and the final product of all this, the capacity to learn from experience—such things did not have to wait for humans and the human brain before appearing on the scene. We thoughtlessly assume they did because, judging by our own psychic experience, we can't imagine it any other way.

"Brainless" is our word for any behavior that strikes us as revealing an unusual lack of intelligence. The word "brainlessness" has a built-in value judgment. As a term of abuse it exaggerates and caricatures a .situation in human psychology. But since we take our experience for the standard of judgment in all matters, nature seems in our eyes condemned to mindlessness—after all we can't find any thinking brains there. Nature's indisputable brainlessness is prematurely equated with the nonexistence of intelligence, imagination, learning capacity, and all the other creative powers that in humans require the presence of an intact nervous system. Our warped perspective has convinced us that our brain generated these powers in the first place, and without it they wouldn't exist.

I am afraid that a substantial part of the wonder nature inspires in us rests on a misunderstanding that has its roots right here. How, we ask, has nature managed to accomplish so much, lacking a brain and all the creative faculties that organ brings us. As if the entire universe, lo these many billions of years, must have had to get along without intelligence, and so forth, because we hadn't arrived yet!

But once again, reality looks quite different from the pictures of it painted by our self-centered way of thinking. We already have proof that brainlessness doesn't necessarily equal lack of intelligence (or, phrasing it more carefully, lack of potential for efficient elaboration of highly complex functional structures): our liver can serve as evidence of that. Modern philosophy of culture teaches us that the "knowledge" ingrained in social behavior patterns can be more intelligent than all the experience stored in individual brains. And a good ten years ago I explained why dealing at length with evolution forces one to recognize how effective a "mind without a brain" can be.[25] This should not be misunderstood as a conclusion to the effect that some supernatural Essence was at work here. It simply points up a fact that, however much it goes against the grain of our imagination, is very real: all the functions and accomplishments implied in terms such as "intelligence" or "understanding" enjoy a quite effective existence outside as well as inside the human brain.

Some of these functions—far more nowadays than many people think—can be carried out by computers. It has long since ceased to be true that a computer can only give out what its human programmer has put into it—modern programming is not at all like this. A chess playing computer, for instance, is fed not data but rules, not prefabricated solutions for specific game situations but criteria and helpful "tips" that enable it to decide on moves in unforeseen (and unforeseeable) situations, after efficiently analyzing the board and thus increasing its chances of winning. The whole operation runs beautifully—and without any brains. The only differences between this and a fully rational performance are, first, an intelligent computer does not (not yet anyhow) consciously experience the feats achieved by its own circuitry and, second, these operations take place outside a living organic brain. To avoid any misunderstanding I hasten to add that one can, of course, always define intelligence (and all other "psychic" operations) so narrowly that only conscious acts carried out by our brain will qualify—and this narrow definition has undeniable practical value. But that's no reason to turn it into a dogma.

Needless to say, the wisdom garnered by whole cultures is genuine knowledge, even though it may not subsist as accumulated, disposable experience in any particular brain. And if F. A. von Hayek rates cultural systems in many cases as "more intelligent" than the individual member of the society in question, he does not mean it in a purely figurative sense.

Learning and intelligence, the search for solutions to problems, decision making against the background of a critical standard that charts the result of past learning processes—all this can be found outside the sphere of consciousness. It all really exists and can have real effects, without being localized in a concrete place (in some brain or computer). It is supra-individual. And there is nothing metaphysical about such a remark. The only thing it contradicts is our old mental habits.

These supra-individual achievements are the key to evolution. First came information gathering, learning by experience, unconscious evaluation and decision making, and *then* came brains. The functions that we habitually label "psychic," because we have conscious, personal experience of them, are older than our brains. They performed their task in the unimaginably long eons before brains (and consciousness) developed. Like everything else, brains could be produced by evolution only because "psychic" functions were directing evolution right from the start.

Our brain, then, invented neither learning nor memory. It merely puts these and related strategies at the individual's disposal. We have to get used to thinking of our brain as the organ with whose help evolution managed to put its own inherent capacities at the service of the individual organism. But this gift, for all the time we have had it, has undergone only a rudimentary development so far. No human being could take charge of his own liver or build a single cell. All our most strenuous efforts notwithstanding we can not understand, let alone copy, more than a tiny part of what evolution succeeded in producing without a brain. We are not, as we tacitly and naively assume and after our dramatic late arrival on the scene, the sole guardians of Mind in all of nature. We are a product of evolutionary history, nothing more, in fact, than a first pale reflection of the principles that have brought into existence everything which we call our "world."

But where do all these powers come from, if they didn't originate in our brains? How did they come into the world? The most likely answer is, they were always there, part and parcel of this world from the first, like elementary particles, physical constants, and the laws of nature. They are inexplicable features of a cosmos that never would have gotten beyond the chaos of the Big Bang, if an orderly pattern of succession had not emerged from the mysterious skein of laws, constants, and other fixed circumstances.

Even the chaos of the Big Bang was only seemingly chaotic, because it is certain that the cosmos issued from it with various features which resulted in the development of increasingly advanced structures. These obviously included the strategies guiding evolution toward the formation of living organisms and, beyond that, of brains, which made those strategies available to the individual.

Those are the facts of the matter—up to this point we have not had any need of metaphysics. But one can't go from the realm of matter and space to an understanding of mind without some sort of leap—even the most spectacular progress in scientific research can't bridge that gap. Bloch made this clear once and for all when he pointed out that even if we could go on a personal tour of inspection in our brain as if it were a workshop, if we could examine and comprehend all its controlling elements down to the last molecule, even then we would not be a whit closer to the mystery of consciousness.

But how are we to understand the empirical fact that our consciousness is dependent upon our brain? If it's basically impossible to give a physiological answer to the question, why did psyche appear in the course of evolution? then what other answer could one give?

I would say that we have to regard the existence of consciousness as a result of the fact that we no longer belong exclusively to the three-dimensional world of ordinary experience, because the human race, after dragging itself along on its evolutionary trek for whole geological epochs, is now once again about to cross an ontological frontier. To be sure, the transition to the next higher level will proceed at the snail's pace that characterizes evolution and that strikes impatient humans as so painfully slow. Still the evidence suggests that this transition has already begun.

When we compared the modes of existence of creatures at

various levels of development, it struck us as perfectly obvious that there was a fundamental ontological difference between them. Between the world of the tick and that of the hen we could find no immediate connection except that both were part of the evolutionary process. One thing we were compelled to realize was that the ontological level we ourselves occupy could not possibly be the last or the highest.

The barrier separating us from the next highest level is not, we argued, without a few chinks. In looking into gravitation or the particle-wave dualism we discovered phenomena whose hidden causes must lie beyond the world defined by our cognitive horizon. We saw that the undeniable order we find all around us could not be explained simply in terms of our world but rather as the reflection of a transcendental structure encompassing that world.

Among the factors that constitute and actualize the order of our world must be counted, apart from natural laws and constants, the processes that reveal "intelligence," even though they take place unconsciously and apart from brains. In this form they played a decisive part, from very early on, in determining the path of evolution, and they contributed to the eventual formation of brains.

In the brain they were, so to speak, condensed and as a result became an active force no longer limited to supra-individual evolution, but for the first time in the history of our planet a source of energy, however inchoate, in the behavior of individuals.

Along with it individuals simultaneously acquired the capacity to become conscious of themselves and the world. Does this not suggest the possibility that just as these functions, which are grounded in a transcendental reality, established themselves through evolution in the heads of individual humans, so other features of that reality, hitherto "beyond" us, may also have begun to spring into life there?

The fact that we now think consciously about the world puts it at a certain distance from us. Does this human achievement mean that evolution is now on the verge of raising us to the next highest level?

If there is a kernel of truth in these paraphrases of something that cannot be directly expressed in language, then perhaps we could compare the relationship between mind and matter, brain and consciousness, to that between light and a mirror. In empty space light remains invisible. It shines only when it meets a surface capable of reflecting it. But however brightly a mirror shines, in no case does it generate the light that it radiates.

Evolution, we said, opens up ever broader domains of transcendence to its creatures. The brain does not produce mind; mind emerges in our consciousness by means of that organ. Psychic reality, the fact that mind exists, which can in no way be derived from the laws of our material reality, might have come about through evolution's carrying our brain to a stage of development that allows the first glimmers of transcendental mind to penetrate within it.

20. The Cosmic Framework

IT IS unthinkable that everything said in this book might hold true only for us here on earth. It would be more than pathological self-centeredness if we were to assume in all seriousness that the leading features of cosmic evolution—its progression from simple to increasingly complex structures, the inevitable transition from these structures to living organisms, and finally the emergence of psychic phenomena—if we were seriously to believe that all these characteristic features of evolution have left their mark on one single place—our planet.

In the final analysis such belief makes us the intellectual center of the entire universe. After centuries of violent controversy we have at last grudgingly accepted the fact that neither the earth nor even the sun is the center of the cosmos. Our ancestors fought this realization until, literally, the blood flowed. We gave in, but only when the piles of proof blocked all escape routes. Nevertheless, to this day most people honestly believe that our earth is still the center of the universe, but in a metaphorical (and far more important) sense, namely as the only place in the entire cosmos where evolution has brought forth life and consciousness.

Wouldn't it be a sign of intellectual maturity on our part to recognize this as an old prejudice in modern disguise? Couldn't we manage, just this once, to show a little self-critical sense and repudiate the error, before we get hit on the head by tangible evidence?

The tremendous popularity of science fiction, even if most of it is unspeakably silly hackwork, may seem to argue against my point here, but it doesn't. Though the books and films in this genre swarm with extraterrestrial visions, that's not symptomatic of any thoughtful challenge to our position in the universe. With a few laudable exceptions the bulk of science fiction simply exploits a scenario whose marketability wasn't discovered until relatively late in the day.[26] Nor do the fantasies of UFO enthusiasts and related groups demonstrate anything like self-critical rationality. They testify rather to a flight into superstition, as amply evidenced by the aggressively sectarian demeanor of such "believers."

By and large people still need to be convinced how irrational it is to assume that the anthropic principle, which was recognizably present only hours after the world came into existence, never operated anywhere except on earth; or that the organic building blocks of life, the biopolymers (which we have found in certain meteorites and, thanks to radioastronomy, in the depths of space even in other galaxies) were subject to further development, culminating in the onset of biological evolution—but only here on earth. Evolution, which we have discovered on earth because that's where we happen to be, transcends the limits of a single planet. It takes place within a cosmic framework. Where could we find reasons, how could we even imagine them, to justify the supposition that the evolutionary process on earth was essentially different from that everywhere else in the cosmos?

One is almost embarrassed to add lengthy examples on top of logical arguments, but experience proves that it's necessary. And so . . . The sun is a ball of gas about 1.5 million kilometers in diameter. It's so large, in other words, that you would need to hollow out only half of it to accomodate the system formed by the earth and the moon (the diameter of the moon's orbit is around 760,000 kilometers). Some 100 billion of such gas balls—or stars —form our Milky Way. If we assume that only about 6% of them (an extremely conservative estimate) are circled by planets, as is our sun, then we get 6 billion planetary systems in the Milky Way alone. But there are other Milky Ways—in fact, the part of the cosmos observed by astronomers thus far contains more Milky Ways (galaxies) than our Milky Way contains stars.

In a cosmos of such unimaginably vast dimensions could we be the only "payload"? The only meaning, the only self-conscious result of evolution on such an incredible scale?

Of course, not all the 6 billion planetary systems in our Milky Way contain a planet whose surface provides large organic molecules with a suitable environment. In our solar system, as far as we know, this is true for only one planet. Since this is the only case we're familiar with thus far, a considerable amount of uncertainty exists as to whether the earth qualifies as typical or exceptional.

Accordingly, estimates by astronomers of how many inhabited planets there may be in the entire universe differ radically. Still, the cosmos is so vast and the number of its solar systems so tremendous that even the most timid reckoning sounds like an exercise in fantasy. One modern handbook of astronomy puts the number of technological civilizations (altogether comparable to our own) in the Milky Way alone at one million.[27]

In the last few years objections have been raised on various sides against such high figures, leading most writers on the subject, for the time being anyway, to base their calculations on extremely pessimistic assumptions.[28] The sum total of inhabited worlds is important because from it can be extrapolated the average distance between them. And knowledge of these average distances, in turn, helps to estimate the chances of a systematic search for radio signals from an extraterrestrial civilization. Up to certain limits (distances of perhaps 10 to 30 light-years—ridiculously little, considering that the diameter of the Milky Way is about 100,000 light-years) we are already capable of conducting that kind of search with existing radio telescopes. The question is, are there enough inhabited planets to justify the expense of looking for them? The jury is still out on that one.

But the argument I'm concerned with here is unaffected by such uncertainties. Even if the earth were the only inhabited planet in the entire Milky Way, and if this were the normal situation, so that a "typical" spiral nebula contained only one planetary culture (a figure some 100,000 times lower than modern astronomy's most pessimistic estimate), even then we would have more than 100 billion places where the seeds of life were sown —and that's only in the part of the universe within reach of our scientific instruments.

We can take it for granted, then, that we are not a unique case in the cosmos.[29] Neverthless the likelihood is that we creatures of earth are in a class by ourselves. A contradiction? No, so long as we don't get it into our heads that our uniqueness consists in our towering superiority. We can rest assured that we, and all earthly life with us, are unique in the sense that we possess an unmistakable, unrepeatable individuality, even against the overwhelming background of all the abundant life in the universe.

This individuality can be substantiated by reversing an argument often used by authors of intimidating expertise to show that we are the only form of intelligent life in the cosmos. In refuting this argument we shall get still another confirmation of the fact that, quite to the contrary, the cosmos must be overflowing with living forms. And at the same time we shall see why, despite its unimaginably enormous size, the universe is still not big enough to repeat a single one of the experiments with life that it has initiated down through time: there are no cosmic twins.

The argument in question deals with a number we have already discussed (in a quite different context) in Chapter 1: with the gigantic figure 10^{130}, the number of possible ways of apportioning the 20 amino acids that compose cytochrome C among the 104 positions in the enzyme's molecular chain (see the illustration on p. 49a).

Early in the book we learned that 10^{130} was far more than the number of all the atoms in the universe. This came in handy because at the time we were trying to show that all forms of earthly life are interrelated. And the sheer immensity expressed in 10^{130} excluded the possibility that the (almost complete) identity of the amino acid patterns of all the enzyme molecules from any given earthly organism might be explained in any other way (such as pure chance) except kinship.

Now, as it happens, two eminent scientists, the physicist Pas-

cual Jordan and the biologist Jacques Monod, have argued, independently of one another, that the statistically demonstrable
uniqueness of cytochrome C (it could never develop by accident
again) constitutes irrefutable proof that life could never have
arisen anywhere else in the universe except on earth.[30]

The gist of the argument is clear. Cytochrome C is indispensable for oxidative phosphorylation ("inner respiration" in tis-
sue). This function is strictly bound up with the complicated
amino acid pattern present in the molecule. Even trifling mutational modifications generally cause cytochrome C to lose this
function, resulting in death by suffocation.

Cytochrome C is thus absolutely essential for life as we know
it and as we can imagine it. But the probability of this molecule's
ever coming into existence is 1 to 10^{130}, that is, practically nil.
Conclusion: the fact that this enzyme could develop on earth
amounts to such a prodigious "long shot" that we can rule out
its reoccurrence anywhere else in the cosmos, however vast it
may be. The argument becomes even more telling if you consider
that these odds hold not only for cytochrome C but also for the
more than 1000 other enzymes that are just as indispensable for
controlling the metabolism of a living organism, as well as for the
incalculable host of elements need for life.

In his controversial book *Chance and Necessity* Monod maintains
that the radically accidental character marking the origin of its
elements makes earthly life a wholly untypical cosmic phenomenon. Viewed against the background of the entire universe,
therefore, human beings and all other forms of life are merely a
superfluous accident, unconnected with the rest of reality, and
hence meaningless.

As logically compelling as this argument may seem to us, it still
doesn't hold up. Both authors, Jordan and Monod, have overlooked a point that invalidates their objection and its cheerless
corollaries: they have left out of account the *historical character* of
the formation of cytochrome C and all the other building blocks
of life.

Upon closer inspection, Monod's turns out to be a subtle variation on the horde-of-apes theme ("How long would it take a bunch of apes aimlessly pounding on typewriters . . ."), which, as we have seen, is untenable. As far as the possibility of extraterrestrial life goes, Monod's case seems logically unassailable only so long as we regard cytochrome C as a product of pure chance. But that's precisely what it isn't. It's the outcome of a process of historical development.

We described how this enzyme must have evolved over millions of years until through a succession of innumerable steps it developed from what was originally a feeble and ineffective molecule into its ultimate form. This process was anything but arbitrary, nor was it determined solely by chance; it was true evolutionary adaptation. That is, from the variant molecules which emerged through random mutations in the generational sequence of as yet primitive cells certain innovative ones were selected for. In each case these were a little more efficient than the bulk of the molecules that had preceded them and that belonged to a type only sketchily enzymatic at first.

Thus the prototype molecules were most certainly not chains of 104 members. We should think of them rather as relatively short amino acid compounds. At the beginning of time, when they had no superior competitors, their low level of efficiency was enough to start up the process of evolutionary optimization. Everything that later emerged from it, all the complex systems and structures we know today, is the result of this process, *not* of pure chance. For any of the presently existing enzymes to develop again by chance would take longer than the life span of the world. But that's not the way it happened.

On a second point, Monod's argument tacitly assumes that optimal "inner respiration" could be guaranteed only by a molecule with exactly the same structure as that of the enzyme cytochrome C.[31] Now unquestionably this is the only solution we know of. It recurs, as we have noted, in all forms of earthly life and thus proves their kinship. But to say that this is the only possible solution is jumping to conclusions.

Such a claim, though, can't be positively refuted so long as we have no evidence from outside our own biosphere to draw upon. But if evolution is a genuine historical process—and no one can doubt *that*—then we may take it for granted that Monod is wrong. We have already established in another context that as evolution proceeds every realization of a given possibility means the elimination of countless other possibilities. The emergence of each new "blueprint," each concrete reptile or mammal, meant that other blueprints, similar but still different, which likewise had the potential of developing at that same point in evolution, would from then on never have a chance to be actualized.

This holds, of course, not only for animal forms but for everything brought forth by the evolutionary process, including the blueprint of an enzyme. Once an efficient principle for oxidation in the cell was found and had taken its first steps toward improvement, no other enzyme model had the slightest chance. It would have had to prevail against a superior solution already in existence—something evolution simply does not allow.

Thus we have to work on the assumption that things might have turned out differently, the number of potential scenarios was legion. Even if the four billion years of the earth's history were repeated a thousand times, this would not suffice to produce a single one of the living forms we find on earth—and thoughtlessly suppose to be the only possible, the only viable ones.

The makeup of earth's natural forms, both animate and inanimate, seems so favorable to life because evolution has unrelentingly adapted them all, under the threat of extinction, to conditions prevailing on earth (and continuously modified by life itself). But even with these limiting conditions, it's still impossible to get an identical repetition of past history. The abundance and variety of ways of adapting are incalculable.

When we look at the patterns of *human* history, we have no trouble in seeing this point. Let's assume that we've set the clock of time back, not all the way to the birth of the planet, but only as far back as the Stone Age. If everything started afresh then, would it be conceivable that 20,000 years later another dictator named Julius Caesar would be assassinated once again by a former friend of his named Brutus? Or that history could be repeated in a single one of its innumerable details? Yet evolution is history too.

And if all this holds true for the earth, how much more valid must it be for evolution on the cosmic scale? Wherever biopolymers, once having emerged in outer space, may have begun to accumulate on the surface of a planet, they met a different set of initial environmental conditions—which were themselves the result of evolutionary chance, already at work.

It does make sense, no doubt, to consider some absolute "boundary conditions." Biopolymers seem to be the same all over the universe. (This is understandable since the 92 elements whose various affinities brought about the development of these heavy molecular compounds are the building blocks of the entire cosmos.) But this makes water in its liquid form indispensable as a solvent to allow necessary chemical reactions to proceed. The likely range of temperatures, then, will be between 0° C and 100° C. It's possible to think of other such theoretically absolute limits, but it quickly becomes difficult here to distinguish objective reasons justifying them from mere hypotheses based on earthbound prejudices or impressions.

But what of other limits? The mass of a planet is largely a matter of chance and so there is a great deal of individual variation. The gravitation on the planet's surface, however, resulting from its mass, has a crucial influence on the type of organic structures that might develop. The spectral composition of the radiation that the planet receives from its sun likewise varies from case to case. So light-sensitive organs face different adaptive tasks in each situation. (Our eyes sense light in that comparatively narrow band of the solar spectrum which is also capable of penetrating through clouds. Thus clouds, the composition of the earth's atmosphere, and the particular spectral qualities of solar radiation have all helped to determine the structure of our eyes and hence the way we see the world.)

The multiplicity of such individual factors that create varying frameworks for evolution is enormous. But it's greatly exceeded

by the number of random events that likewise direct the path of
evolution and determine which of the countless possibilities that
might be realized at a given moment will in fact be realized—at
the expense of all the others. The beginnings of evolution, then,
may be thought of as roughly similar throughout the universe.
The primal "stuff" was evidently pretty much the same everywhere. But with every step taken by evolution under the isolated
circumstances of each planet, each one of the innumerable cosmic experiments with life must have taken on a steadily more
unique and unmistakable character. Even the cosmos isn't big
enough for historical accidents to be repeated.

But though we can be sure that a plurality of cosmic life forms
exists, we still have to be on our guard against anthropocentrism.
We are most certainly not the apex of the universe. Throughout
evolution the principle of mind, through the mediation of ever
more powerful brains, has found its way into the material world;
and we are most certainly not the final point, the *ne plus ultra* of
this process. Actually, if we recollect how irrational our behavior
still is, how innate instincts and fears still prevent us from doing
what we know is right, when we realize that we are gambling our
existence on the insane arms race—then the contrary seems more
likely. Our species has doubtlessly already entered on the transition from animal to human, but we have just as surely not left all
of the earlier form behind us. The *Homo sapiens* of Enlightenment
philosophers, the creature on whom they based their optimistic
social designs, continues to be a future hope, not a present reality.

In all likelihood there must be numberless places in the universe where evolution has gotten beyond the stage we now find
ourselves at. Since the tempo of evolution also varies from case
to case, the most plausible assumption is that we belong somewhere in the middle ground of cosmic evolution as it currently
stands. This would mean that there are billions of planets with
living beings at a level of development lower than our own, but
at the same time billions of other planets inhabited by beings
unimaginably superior to us.

These creatures would be equipped with brains that would help their owners to a much larger share of that mind which has just now begun to shed some light, though as yet a relatively faint one, on our own heads.[32] Such beings would have a cognitive horizon encompassing parts of the objective world that are beyond our experience. We call such domains of reality "transcendent," because we have no access to them, although we have already discovered that they must be out there.

In this way the abstract-sounding claim that transcendental realms of reality can be integrated into our consciousness is given a concrete content. And the same goes for the statement that over the course of its history evolution continually transforms new areas of the objective world into subjectively experienced reality. This idea doesn't refer merely to a hypothetical possibility or our own past or future. On the contrary, the individually varying *tempi* of evolution throughout the universe would bring about the simultaneous existence of subjective worlds spanning the widest ontological gamut. These worlds would be separated not only by physical distance, which could probably never be bridged, but also by equally insurmountable barriers of the sort that make it impossible for us to convey to a chimpanzee, however intelligent, even the vaguest notion of what we think about the world.

One more aspect of this subject that strikes me as important: epistemology and the study of how cognition evolves have shown us that the specific route taken by biological evolution on earth has to a great extent determined how we see the world. But we are surely not the only ones to whom this rule applies. The inconceivably numerous subjective worlds in the cosmos differ not only in the level of development they have attained; they must also correspond to basically different mental concepts.

Once again we shouldn't let our speculative fantasies run away with us. Here too there are limiting conditions. The matter that intelligent extraterrestrial beings have to deal with is everywhere the same. The natural laws and constants are also identical. Hence the subjective world pictures of such creatures, could we

compare them with ours, would definitely reveal parallels and
perhaps even direct agreement with regard to certain fundamental structures. Anyone who managed to get an overview of all
these subjective worlds would be able to recognize that they
referred to the same cosmos, that they all claimed to be copies
of this one common universe.

None of these world pictures, however, could ever be congru-
ent with any one of the others. It is mathematically impossible.
The unique evolutionary path leading to each one of them is
unrepeatable. And none of them could embrace the cosmos in its
totality. They would all select only a part, large or small and
never the same part, of the whole.

These conclusions may well fit snugly into the picture sketched
out in this book of the evolving world as an act of creation. If
evolution is the mode in which we experience the moment of
creation (still going on), then the question suggests itself, when
might this moment come to an end—or *will* there be a natural
end to evolution?

As I see it, the answer is yes. Of course, we can't be absolutely
sure. We have to speak cautiously because of the limitations
imposed by our cognitive horizon. Still, we can say that evolution
is increasingly opening the door to this world for mind (the
existence of which can't be convincingly deduced from the world
itself). Consequently, evolution could be described as a process
of development in which the cosmos has begun to blend with the
principle of mind, which was the prerequisite of its emergence
and of the order that unfolds in its history. We might say that the
history of the evolving universe is the history of that fusion.

We ourselves experience the outcome of all this as the capacity
for conscious reflection. We have become aware of our own
existence and of the fact that we exist in a cosmos full of unfathomable (for us) enigmas and mysteries. Much to our surprise we
succeeded over the course of time in finding answers to some of
these enigmas. Thus we learned that we are a part of this cosmos,
having emerged from it during a historical process that began at
a point in the unimaginable depths of the past, led all the way up

to us, and will in the future transform ever greater realms of what now lies beyond our reach into experiential reality.

This suggests the possibility that evolutionary history might come to a natural end once it finally produces a consciousness large enough for the truth of the entire universe. If so, the natural end of evolution would occur in that far-off moment when this empirical world and transcendent mind are transfused into one another.

But hopes for such a total eschatological realization of truth in the world cannot, of course, rest solely on human consciousness —not even if we develop to the fullest all the as yet unknown evolutionary possibilities of that consciousness. Till the end of time the pattern of evolution on earth will bear the mark of its peculiar individual history. It may exhaust that aspect of the cosmos which its history assigned to it. It may say all there is to say about it—but only about that one aspect.

And there are many others. At this point in our reflections it looks as if the reality of an incalculable number of different forms of consciousness in the cosmos is not only logically compelling but an existential necessity. We have to accept it. As long as time lasts no single one of these incipient forms will ever flower into a total, all-embracing understanding. Every one of them remains bound to the laws of its own individual history. But might not all of them put together, might not their common effort, prove sufficient at the end of time to close off evolution with an act of knowledge that grasps the entire cosmos?

21. Evolution and Transcendence

NONE OF this, obviously, should be taken literally. Properly understood, the message of this book comes down to the idea that the road to knowing the truth is permanently closed. That is why the end of evolution, the universal doomsday, will differ in some unthinkable way from my tentative description in the last chapter.

In order for our remarks on the end of the world to have any legitimacy, we needed a concept of the horizon that in this final stage of the cosmos evolution will have thrown open to knowledge. As far as the moment of evolution goes, to which we are bound for all time through our earthly existence, this possibility of a triumphant evolutionary climax remains a utopian notion.

For this reason this book should be read as a series of speculative sketches. These do not serve to make an objective record of verifiable facts, they are not scientific statements, and they do not observe the limits of the positivistic method. Convinced as we were that the framework of positivism was too narrow for the full range of reality, we decided to go beyond it and disregard its gag order.

We had to pay a price for this infraction, but it didn't seem too high: we had to give up strictly literal discourse.[33] This is no great sacrifice once one realizes how readily we have always made it in all other domains of our thought and experience. Not only do all the arts and sciences—including the most exact science, physics —deal with specific experiences that can only be paraphrased in a metaphorical, image-filled language, but all of us constantly and unabashedly make use, not of the literal, but the figurative meaning of expressions, as soon as what we have to say passes beyond the narrow realm of immediate interpersonal relations.

Just because something is put in mythological language, therefore, does not mean it is false. Much, in fact most, of what we "know" cannot be expressed any other way. One of the Enlightenment's most aggravating errors was its idea that mythic statements do not refer to the real world, that they are an archaic and hence outdated form of speech, *ipso facto* false or meaningless. If this were actually the case, we should have to question the reality of the objects described in the metaphorical language of nuclear physics—a conclusion few nowadays would endorse.

The ultimate reason for all such misunderstandings is always the same: our uncritical tendency to consider reality of which we have "concrete" experience to be reality *par excellence*. In one quite specific sense this is undeniably right: what we ordinarily call "the world" differs for us from all other realities, whether merely conceivable or actually existing anywhere else, in that it is the reality to which we are tied down by the accidents of our individual history, by our place in the universe, and the cosmic moment of our existence. But the criteria that result from these conditions are exclusively subjective. They hold good without exception, in exactly the same way and with the same authority, for *every* relationship between a subject and his or her world. The number of such cases, as we have seen, is infinite.

Our "world" enjoys its special status with us simply because it represents the part of reality that our intellectual capacity, stamped as it is by the random events of our evolutionary history, selects (in the most arbitrary, self-willed fashion) from the sum of objective reality. There can be no doubt, we noted, that the slice of reality with which we have to make do is, of necessity, infinitesimally small. The other side of this fact is our discovery that our old familiar world, apparently so tangible and real, suddenly begins to be extremely problematic and mysterious, inexplicable on its own terms, and finally a mere hypothesis, the moment one thinks to take a few critical pokes at its seeming solidity.

All these reasons make it unacceptable to discredit areas of reality inaccessible to our experience by pointing to the sensory immediacy of our everyday world. Transcendental reality is every bit as real as the world of our experience.

In the context of social structures, artistic developments, or political decisions we may look upon ourselves as the measure of all things. But vis-à-vis the objective conditions of our existence we are no such thing. We are only a part of the whole, part of that all-encompassing history that produced the entire universe. The intellectual capacity that has accrued to us so far turns out to be hopelessly inadequate when we seek for the truth about this world. But it's enough to let us see that the world doesn't end where our intelligence runs up against its limits. "And for what may be beyond, the eyesight of Lilith is too short. It is enough that there is a beyond."[34]

From this standpoint mythological statements and metaphysical speculations lose the stigma of irrationality that still attaches to them in the view of many people. Instead they prove to be the consequence of a rational, self-critical examination of the objective conditions of one's own existence. Not rationality, but only the naiveté of unreflective realism could lead us to treat as a phantasm the domains of reality lying outside the range of human reason—which has already recognized its limitations.

On the other hand, we would be lost in the mist of countless imaginary possibilities, if we misunderstood this argument as a license to contrive whatever world picture we please. Plausibility and freedom from contradiction do not suffice. In the territory of metaphysics the number of legitimate paths is endless. Every step here carries with it the danger either of slipping into noncommitment or of getting stuck in superstitious concreteness.

But here too, though some refuse to heed them, there are guideposts and landmarks. The most important of these, to repeat, are first, the abandonment of all literal interpretations of the language used to describe the path and, second, the categorical rule that none of these descriptions in any detail may contradict our knowledge of accessible reality. However meager and incomplete it may be, this knowledge is all we have to hold onto, unless we are willing to stop believing that our thoughts make sense.

That is why it is justifiable to speak of a Day of Judgment. The mythic prediction of the coming end of all time is legitimate because it does not contradict what we know about the history of the cosmos. Despite all the gaps in this knowledge, we can be certain that cosmic evolution will come to an end, just as it demonstrably had a beginning.[35]

Now one might think that this end will be arbitrarily determined. Thus history might not actually come to an end, but simply stop, without regard to evolution, to which it was previously identical. Again there is no refuting the possibility that the cosmos, which was unfolding into a pattern of order too vast for us to conceive, might prove in its last moment to be meaningless after all. In that case everything that it had brought forth up to that point would be retroactively undone.

But again we may ask a counterquestion: would it still be reasonable to press our skepticism so far as to maintain that an order which has steadily developed over a dozen billion years is a product of pure chance? Would it be plausible to assume that the mind-boggling outlay represented by the reality of cosmic history could mean "nothing"? The question cannot be settled either way, but the risk in making a personal decision seems slight.

Hence we have a right to trust that cosmic history won't simply break off but will rather find "its" end. And since in what we can see of evolution we observe a continual expansion of mind, we can think of the end, the "Judgment Day" of history as that future moment in which mind will have absorbed this world into itself. "The day will come when there will no longer be man, but only thought."[36]

But at the same time we have to acknowledge once again the wisdom of cultural tradition, which so mysteriously surpasses all the capacities of individual thinkers. Long before we stumbled onto the discovery that the end of time had to coincide with an evolutionary summit which would have nothing in common with our ideas about it, tradition was already warning us not to speak of this event except in the indirect language of images.

This doesn't mean that our expectation of there being only thought at the end of time is false. It simply says that the reality of that moment will outdo and render irrelevant whatever we can say about it today with the means at our disposal.

Scientific and religious statements fit seamlessly together here. More than that, they complete and confirm one another—which should not surprise us. There is only one cosmos (which we must simultaneously understand as creation), and there can be only one truth—a truth comprising both these aspects. The fact that we are capable of fully grasping neither the cosmos nor creation nor the truth encompassing them both in no way changes any of this.

Some people have the impression that what science tells us about this one cosmos is incompatible with what religion tells us about creation, but this can only happen when the rules mentioned above are neglected. From the very beginning the literal meaning of the mythical language used by theologians to transmit their message had very little to do with the content of that message. This was true even in those days two thousand years ago when the images characteristic of that language first appeared as an expression of living faith.

Since it was not the literal meaning that mattered, there was no need at the time for explicit interpretation. All of Jesus' contemporaries, from the scribes to the tax collectors, were intimately aware of the indirect, metaphorical meanings of his message. But that was two thousand years ago. The cultural milieu of Christ's time, its world picture, has long since vanished, and along with it so have the semantic "overtones" of the mythological language spoken then. Gone are the associations, the echoes, the reminiscences, and all the other modes of communication that transcend the literal meaning of the text—and without which we cannot really understand what the authors of those ancient texts wished to say to their audience.

What we have today is only the skeleton, the bare bones of words and sentences. We can always translate these into the language of our time, and as a vestige of the the period they derive from they awaken our respect and reverence. But the gamut of connotations, the depth of meaning once tied in with them has been lost. And, as we have found, once mythical statements are reduced to their bare literal sense, they degenerate into superstition.

If we are honest with ourselves, we can't deny that we have by and large succumbed to that danger. From Christ's ascent into heaven (viewed as a literal flight into space) to the quite corporeal glorified body, blessed with magical powers; from belief in suspension of natural laws (the only convincing proof, some think, for God's omnipotence) to the assumption that one can intervene telekinetically in the world by concentrating one's mind through prayer[37]—the opportunities for superstitious misunderstanding of the old texts are everywhere. We meet this sort of thing by no means only among the "simple souls," but also, and with distressing frequency, among the representatives of the Church. And in that case collisions between religion and science are unavoidable.

But once the conflict has broken out, its real causes are quickly lost sight of. Then the chief concern of both sides is that any yielding, even so far as to take the other's argument seriously, might be read as a surrender of part of one's convictions, of the partial truth one side monopolizes for itself. Fear of this misunderstanding all too readily drives both parties to take refuge in the claim to be sole distributors of the truth.

At least it has been that way in the past. For generations now theologians have felt surrounded by the forces of science, which they see as undermining the foundations of religious faith, as "materialistic" and hostile to all things spiritual. And for their part scientists have fallen prey to the enticements of a narrow positivistic definition of truth, which was tailored to fit their disciplines and which made theological statements sound to them like so much empty verbiage.

Lately the battlefield has grown quiet. The time of furious attacks and mutual defamation has come and gone, giving way to peaceful coexistence. But it's still too soon to talk about a religious-scientific consensus. The doctrine of the "two truths" is indicative of a stalemate.

The "open" discussions and "frank" exchanges between scientists and theologians, a common occurrence these days, are especially noteworthy for their studied politeness. Both groups have agreed . . . to be nice to each other and to avoid issues that might cause embarrassment. But this peace is a sham, because it tolerates the psychic split people must live with unless they are completely satisfied with one of the two versions of the truth.

We have had so much time to get used to this split that it hardly bothers us any more. Nowadays those who want to believe without ignoring the findings of science are forced to make themselves at home in a world whose two halves don't fit together. This sort of philosophical schizophrenia hampers free intellectual development and warps personal behavior. Most people affected by it may seem to be managing well enough. The human psyche is, after all, the most adaptable organ on this planet. But others take a more honest approach to the problem by settling for one of the two truths. They withdraw in resignation to an atheistic standpoint or to its counterpart, a one-sided creationism that rejects scientific knowledge as a diabolical illusion.

But perhaps there is another way out of this situation, a bridge to span this mental split. Perhaps—and this book is an attempt to call attention to such a possibility—the two halves of our world picture can be put back together again, if we remember the rules governing mythical statements; if we can get up the courage to distinguish between the actual core of religious tradition and its linguistic sheath. It will take a little nerve to consider that literal belief in scripture might be less an expression of unswerving loyalty and more an unconscious fear of risks; that the obligation to preserve tradition threatens to betray that tradition when it begins to stifle its vitality.

This is the danger that menaces religious tradition in modern society. The content of religion is being buried by the sediment of an archaic language, no longer alive and therefore no longer really comprehensible in its original sense. The power of the mythic images in which the message of religion has come down to us has so diminished that they fail to grip us with the power needed to understand their true meaning.

The obvious solution to this dilemma is to bring in a new language. The gospel has to be given a new mythical envelope, it has to be translated into a new metaphorical language as powerful, as up-to-date, as broadly meaningful as the ancient texts once were for those who used them two thousand years ago. But we need not go out and invent this new language; it already exists. It is, as we have seen in some detail, the language with which contemporary science describes the cosmos.

At the heart of this description lies the concept of evolution. This will surprise no one except a person who thinks, as most laypeople do, that evolution is a subsection of theoretical biology. In reality the word relates to a central element in the scientific world picture that has had a decisive influence on the way modern humans views themselves. All that we have learned about organic forms over the last 200 years, especially Darwin's grand synthesis, constitutes a principle that stretches far beyond the realm of biology. Evolution is, in the truest sense of the word, everything, that is to say, an event that encompasses the entire universe. Evolution is synonymous with the absolutely revolutionary discovery that the cosmos, the world itself is a historical process.

It's surprising how close one comes to age-old theological positions once one turns to the language of evolution for help in formulating a religious message. All one need do is try to describe fundamental theological positions not in terms of a static, inalterable, "as it was in the beginning, now, and ever shall be" world, but against the background of a cosmic development spanning all time. Anyone making this attempt will consistently get the impression that the old formulas are awakening to a radiant new life after being transplanted in the fresh soil of a language that addresses our contemporary understanding of the world.

One of the examples I cited to illustrate this was the notion of the "Beyond." This term has a significance for religion (or at least for religions consisting of more than a set of ethical rules) that is akin to the significance of the term evolution for the scientific world view. The essence of religion is its belief in the reality of a Beyond that embraces and transcends all of human experience.

But this key religious concept necessarily lost much of its persuasive power in the context of a static world picture. It fell into jeopardy the moment the possibility arose that people might succeed in understanding the world. In the face of a self-contained world that obeyed rationally intelligible laws, the Beyond inevitably began to sound more and more like a figure of speech. Where could this Beyond actually exist?

The further human reason thrust its way into the depths of this world, the smaller grew the space available for theologians to lodge their heaven. The biologist Ernst Haeckel spoke mordantly of "God's housing shortage," because the theologians' unwavering claim that the Kingdom of God lay "beyond" this world seemed, when voiced in the context of a closed universe, to point to a place for which there was, in every sense of the word, no more room.

But in a world that is still becoming, still moving toward completion by means of evolution, things are entirely different. The fact of evolution—supposedly so hostile to religion[38]—has shown us that reality doesn't end where our experience stops. Not philosophy, not classical epistemology, but evolution compels us to recognize an "immanent transcendence" that immeasurably surpasses our present cognitive horizon.

Let me repeat, this transcendence cannot be immediately equated with the theological Beyond. But its discovery creates something like an open door in a world that had seemed until then to be ruthlessly closed to any such possibility. Beyond this opening we can glimpse an ontological progression of ever more fully developed levels of knowledge. And no one could contradict us, should we wish to think of the final level as "heaven," where, religion tells us, lies the key to the meaning of our imperfect world.

NOTES

1. See *Im Anfang war der Wasserstoff*, 4th ed. (Hamburg: 1975), pp. 19–51.
2. Friedrich Nietzsche, *Thus Spake Zarathustra*, part III.
3. C. F. von Weizsäcker, "Astronomie unseres Jahrhunderts," 9th lecture in the series, *Die Tragweite der Wissenschaft* (Stuttgart: 1964).
4. See Donald T. Campbell, "Downward Causation in Hierarchically Organized Biological Systems," *Studies in the Philosophy of Biology* (Berkeley: 1974).
5. Evolution can only develop material that is already there; it can never (not on earth anyway) make a free and fresh start. The more complex, elaborate and functionally specific a genetic blueprint is, the slimmer are the chances for fundamental change.
6. Paul Davies, *The Runaway Universe* (London: 1978), p. 185.
7. B. J. Carr and M. J. Rees, "The Anthropic Principle and the Structure of the Physical World," *Nature* 278 (1979), 605–612.
8. J. A. Wheeler, as quoted by Carr and Rees in "The Anthropic Principle."
9. Jacques Monod, *Chance and Necessity* (New York: Knopf, 1971).
10. This corresponds to the so-called Equivalency (or Identity) hypothesis, which is yet another response to the mind–matter problem. See Gerhard Vollmer's essay, "Evolutionäre Erkenntnistheorie und Leib-Seele-Problem," *Herrenalber Texte* no. 23 (Karlsruhe: 1980).
11. For a particularly crass example of this see B. F. Skinner's *Beyond Freedom and Dignity* (New York: Knopf, 1971).
12. Ernst Bloch, *Das Materialismusproblem, seine Geschichte und Substanz*, vol. 7 of his *Collected Works* (Frankfurt am Main: 1972), p. 289.
13. The best known example is Friedrich Engels, *Anti-Dühring*, vol. 20 of the *Collected Works of Karl Marx and Friedrich Engels* (East Berlin: 1973).
14. C. F. von Weizsäcker, *Die Einheit der Natur*, 4th ed. (Munich: 1972), p. 289; in English translation, *The Unity of Nature* (New York: Farrar Straus Giroux, 1981).
15. Konrad Lorenz, *Die Rückseite des Spiegels* (Munich: 1973), pp. 48 ff.; in English translation, *Behind the Mirror*, trans. Ronald Taylor (New York: Harcourt Brace Jovanovich, 1978).
16. Gerhard Vollmer has argued against me on this point, objecting that such a sharp distinction as I have drawn does not actually exist. Future research, he says, will most likely succeed in breaking down the all-inclusive concept of "mind" into partial functions and will then work out specific criteria for each of these. "Some of these capabilities go way back into the evolutionary past, others are found only in vertebrates, and others again only among primates or in man." Thus the development of "mind" (in general) could be analyti-

cally divided into a plurality of small steps, each one of which could be viewed as a "fulguration" in its own right.

I don't find this objection convincing for two reasons. First, when I talk about "mind" or "the psychic," I am *not* referring to the functions (or anything like them) that Vollmer mentions, such as memory, abstraction, power of speech, and so forth. Someday computers will be able to perform all these operations (insofar as they can't do so already) *without any consciousness.* And, as will be argued later in the text, such functions have already controlled by far the greater part of evolution up to now, once again unconsciously. What I mean is the phenomenon of *subjective experience,* which is fundamentally independent of Vollmer's partial functions, although it does, for reasons as yet unknown, sometimes accompany them. As I see it, this phenomenon did not come into being in accordance with the "all-or-nothing principle" that typifies a "fulguration."

Furthermore it strikes me as a flimsy *ad hoc* argument to claim that in the future (!) it will be possible to break down mind into smaller and smaller evolutionary steps until each individual one can be construed as a fulguration. This method could also be used to treat, for example, embryonic development as a chain of tiny fulgurations, which would simply not do justice to this process. See my article, "Gedanken zum Leib-Seele-Problem aus naturwissenschaftlicher Sicht," *Freiburger Universitätsblätter* 62 (1978), pp. 25–37; Gerhard Vollmer, Evolutionäre Erkenntnistheorie und Leib-Seele-Problem," *Herrenalber Texte* no. 23 (Karlsruhe: 1980).

17. Both this part of the argument and the quotation from Ryle come from Hans Sachsse, "Wie entsteht der Geist?" *Herrenalber Texte* no. 23 (Karlsruhe: 1980), pp. 91–105.

18. Erwin Schrödinger, *Geist und Materie* (Braunschweig: 1959).

19. Gothard Günther, *Das Bewusstsein der Maschinen: Eine Metaphysik der Kybernetik* (Baden-Baden: 1963).

20. Michael S. Gazzaniga, "The Split Brain in Man," *Scientific American* (Aug., 1967), p. 24.

21. Bloch, *Das Materialismusproblem, seine Geschichte und Substanz,* pp. 311–312.

22. Oswald Kroh, "Das Leib-Seele-Problem in entwicklungspsychologischer Sicht," *Studium Generale* 9 (1956), p. 249. In a similar vein see J.B. Best, "Protopsychology," *Scientific American* 208 (1963), p. 55.

23. Gerhard Vollmer, "Evolutionäre Erkenntnistheorie und Leib-Seele-Problem, pp. 14 ff.

24. Konrad Lorenz, "Gestaltwahrnehmung als Quelle wissenschaftlicher Erkenntnis," *Zeitschrift für experimentale und angewandte Psychologie* vol. 6, no. 1 (1959), p. 127.

25. See my *Im Anfang war der Wasserstoff,* 1st ed. (Hamburg: 1972), pp. 238–251.

26. The authors of these productions, as a rule, waste no time reflecting on what the existence of nonhuman forms of consciousness might mean. For the most part they simply transfer old-fashioned crime or adventure stories to highly exotic "extraterrestrial" settings.

27. Karl Schaifers and Gerhard Traving, *Meyers Handbuch über das Weltall,* 5th ed. (Mannheim: 1972), pp. 699 ff.

28. S. von Hoerner, "Where Is Everybody?" *Naturwissenschaften* (Sept. 6, 1978).

29. Can we exclude the possibility that belief in the existence of angels might be

a kind of insight into the unlikelihood of our uniqueness, a hint of the existence of beings unimaginably superior to us?

30. Pascual Jordan in *Sind wir allein im Kosmos?* (Munich: 1970), pp. 151–165; Jacques Monod, *Chance and Necessity.*

31. We may recall here that every specialization leads to a narrowing of the possibilities for development (Heinrich K. Erben's "canalization effect"). Thus the evolutionary path leading to the optimization of cytochrome C's enzymatic effectiveness quickly grew ever more narrow very early on. With every single improvement the number of possible mutations that might have produced a further enhancement of performance steadily shrank. Thus as far back as a billion years ago this molecule had become so successfully specialized that practically any alteration (in its functionally important sections) could have only led to setbacks.

Thus it's quite correct to say that in its current form cytochrome C is indispensable. But that doesn't mean that, for instance, had evolution approached the problem of metabolism from a different angle an entirely different solution might not have been found.

32. An argument can be made that biological evolution might end as soon as its products (ourselves) acquired sufficiently complex cybernetic structures, enabling them to evolve autonomously, without the help of organic, "living" technicians. See Werner Kreutz, "Das Geist-Materie-Problem aus naturwissenschaftlicher Sicht," *Herrenalber Texte* no. 23 (Karlsruhe: 1980), pp. 61–72.

33. On p. 175 I wrote that I meant my idea of evolution as the moment of creation literally. This was simply asking readers to take this idea *no less seriously* than all the other notions we entertained in the context of creation (all of which, naturally, could be formulated only in the language of myth and to that extent may *not* be understood literally).

34. George Bernard Shaw puts these words in the mouth of Lilith in *As Far as Thought Can Reach,* Part 5 of *Back to Methusaleh* (*Collected Works,* vol. 16 [New York: 1930], p. 262.

35. See Part I, note 12.

36. Shaw, *Back to Methusaleh,* p. 262. Shaw adds here, "And that will be eternal life."

37. The more directly the person praying aims at fulfilling a concrete wish, the more open that person lays him- or herself open to the charge voiced by Feuerbach, Freud, and other critics of religion that prayer simply expresses an infantile projection. On the other hand to the degree that prayer is the expression of the search for a reality transcending our world, this objection misses the point. "To pray is to think about the meaning of life." (Ludwig Wittgenstein, *Diaries,* quoted in Küng, *Does God Exist?* p. 506.)

38. Why was Sigmund Freud, who after all taught that belief in God was only a form of "infantile wish fulfillment," never attacked by the champions of orthodoxy with anything resembling the vehemence directed at the founder of the theory of evolution? Might one not conclude from this that for many of those critics it was more objectionable to consider people related to animals than to consider God a sheer illusion?

Index